Super-Matter

Functional Richness in an Expanding Universe

Pravir Malik, Ph.D.

ISBN-13: 978-0-9903574-6-9

To Sri Aurobindo

Author's Note

We live at an incredible time in an incredible universe.

Through a play of Light its codified properties project out to create layer upon layer of fourfold complexity. Space, Time, Gravity, and Energy are perhaps the first layer of Light's codified properties and emerge as fundamental to cosmic dynamics to-date. Fields such as the electromagnetic field form a second layer. Then quantum particles, atoms, cells, and all the complexity of Natural and Human history form subsequent layers, all the while continuing to materialize the inherent fourfold-richness inherent in Light.

The fate of such a continued fourfold development leading to a foundation of super-matter now lies in our hands. Perfectly positioned between two apparent extremes, the human lies between approximately 30 orders of magnitude of micro- and macrocosm on either side that they are an intimate part of. All the depth and mystery of the microcosm, starting from the quantum-levels lies already implicit within the human. Perhaps even it can be said that the human is the result of all the emergent dynamics of these 30 orders of magnitude.

But now the scales of causality are reversing and the question is whether the human can similarly shape the further development of the macrocosm perhaps through mastery of the microcosm. Such audacity is only possible because all is Light and all is of Light. There is an intimate and implicit connection between apparent separativity regardless of scale and for the first time in history Nature appears to have strongly separated levers such as Will, awareness, and complex thought, amongst other faculties to a creature, and to exercise these to intimately structure and become a living part of the macrocosm is the challenge that lies before us.

The scales of a quantum-cosmic balance, therefore, depend on the extent to which a continued and cohesive space-time-energy-gravity quantization, fuelled by such will and other human faculties becomes operative. Such quantization spurred by the need for functional-richness results in super-matter. But further, such quantization may be intimately related to the energy known by cosmologists as 'dark energy' that continues to fuel the expansion of the universe.

In such an expansion the once recurring Big Bang, sign of the tail-biting Ouroboros in the throes of a constant cycle of destruction and rebirth, potentially yields to a new series of inter-galactic-bangs or other such creative mechanism, and sets into motion an endless development with super-matter building on super-matter.

Pravir Malik,
San Francisco

Previous Books in *Cosmology of Light* Series

- Book 1 – A Story of Light: A Simple Exploration of the Creation and Dynamics of This Universe and Others
- Book 2 – Oceans of Innovation: The Mathematical Heart of Complex Systems
- Book 3 – Emergence: A Mathematical Journey from the Big Bang to Sustainable Global Civilization
- Book 4 – Quantum Certainty: A Mathematics of Natural and Sustainable Human History

SECTION 1: LIGHT, A BIG BANG, THE SEED-EQUATION, & QUANTA

This section explores a theory for the creation of quanta related to the speed of light. Initially quanta and a seed-equation are created in the Big Bang. Quanta are intimately associated with the seed-equation and acts as a doorway, as it were, into the multi-layered structure represented by the seed-equation. But time, space, and gravity can also be thought of as existing through quantization. Ability to effect such quantization through an urge for functional-richness can result in super-matter. Such quantization can also be linked to expansion of space. Hence this section explores:

- Light @c, Quanta, & The Reality it Creates
- Light @Infinite
- The Big Bang & The Seed-Equation
- What is Light & What are Quanta?
- Structured Time, Structured Space, and Their Relation to Quanta
- Quantization of Space, Time, Matter, and Gravity
- Functional-richness, Super-Matter, and expanding Space

Chapter 1.1: Light@c, Quanta, and The Reality It Creates

Everything is made from light. But how can this be. How can the dense material objects that make up our world – the stones, cars, buildings – be made from light? How can our thoughts, our emotions, our cells and bodies be made from light? How can all the animals and plants and clouds be made from light? Here is an exploration on now this may be possible.

To begin this exploration think for a minute as to how fast light is traveling. If we turn on a light switch in a room it immediately gets filled with light. If we turn on a light torch on a dark night it immediately lights up a path through the darkness. And yet the light from the sun takes about eight minutes to reach the earth – this speed through a vacuum is referred to as the physical constant 'c'. So while light is moving incredibly fast it is still not moving at the infinite speed we may think it is moving at.

Now, it is because the light is not traveling at an infinite speed that quanta and subsequently the world of matter can appear. All matter is made from atoms. So let us enter into the world of a hypothetical atom in the process of creation to explore how this might be. Imagine that some form of light originates from what will become the nucleus within an atom. By virtue of its finite speed light would take some time, however small, to travel any minute distance within the forming atom. During that fraction of time there is a build up of energy and it is this build up of energy that forms a packet or quanta, as it were, that allows matter to form. In the absence of such a build-up potential-matter would simply disperse. Hence, it is this build up on energy that variously expresses itself as quarks and leptons and bosons and subsequently atoms – the very building blocks of everything that is.

So, in this view, it is because of the finite speed of light, c, that matter is created. That is not to say that matter will be created wherever light is traveling at c, but the possibility exists, so long as other conditions are also fulfilled. But let us leave aside the other conditions and the exact process of such a creation for now.

But what else is implicit or made possible because of the finite speed of light? Well, imagine traveling on a ray of light from the sun to the earth. Imagine that you are in minute 4 of the approximately 8 minute journey. As you look back you will see that 4 minutes in the past you were at the sun. 4 minutes in the future you will be at the earth. And in the present moment you are somewhere between the sun and the earth. So this limited speed of light already creates the concept of time and specifically of the past, the present, and the future.

So, four incredibly fundamental things are created because of the finite speed of light: quanta and therefore matter, the past, the present, and the future. It could be said, therefore, that this matter-based time-bounded universe is a result of the finite speed of light.

Looking at this in another way, it is known too that in a denser medium the speed of light further slows down. Hence, c becomes a fraction of c when traveling through water for instance. So by reversing the logic of such a process this may show that if light slows down from an infinite speed to some other lesser speed such as a multiple or a fraction of c, then the material reality will have to alter, in a similar way as the material reality between vacuum and water is different.

Chapter 1.2: Light @Infinity

But what if light can be experienced at other speeds? What if light can travel at other finite or an infinite speed beyond the c we experience it at, which by the way is an astronomical 186,000 miles per second in a vacuum?

Let us imagine for a minute that light can travel infinitely fast. Think about a big area or volume with a light source at the center. Now since the light travels infinitely fast it will fill up the entire volume instantly. This will be true no matter how large the volume is – it could be the entire solar system, an entire galaxy, or an entire universe. So that light is going to be instantaneously present everywhere – it is going to be omnipresent.

A further thought experiment may show what such omnipresence may look at by considering the night sky. It is because of the finite speed of light that the night sky has only spots of light, however many, across it. But if light were traveling infinitely fast then the night sky would be a canvas or a sea of brilliant white since the light from every corner of the universe, wherever a light source or star existed, would immediately be present. It would appear, in such a thought experiment, that we existed in a sea of light.

Now, since the light is already present everywhere in whatever volume, that is, since that light has already filled up the entire volume, there is nothing else that can similarly arise there that is not of the nature of light. Even if something else were to arise, being surrounded by light it would eventually succumb to that light. So the light is all-powerful or omnipotent within that volume.

Further, since the light exists simultaneously everywhere in that space and has a complete knowledge of itself it therefore has a complete knowledge of that space or of anything that can arise in that space. So it is all-knowing or omniscient in that space.

Finally, the light connects everything together instantaneously and holds these connections and the things connected in its nature, so it is all-nurturing or omninurturing within that space.

So, as a result of the infinite speed of light, light appears to have properties of omnipresence, omnipotence, omniscience, and omninurturing.

Chapter 1.3: The Big Bang & The Seed-Equation

We hear about the Big Bang as the start of the universe. In this Big Bang matter is created. But in the examinations just presented the creation of matter is nothing other than the result of light traveling at the finite speed, and in this universe at 186,000 miles per second in vacuum. So we can say that the Big Bang, the apparent start of the universe, is the result of a slowing down of light from an infinite speed to some finite speed.

When light slows down then energy accumulates in packets or quanta and this results in what was inexpressible being able to express itself as matter. This can be thought of in terms of an incredibly rapidly moving stream of water. If the water is traveling so fast then no boundary will be able to contain it and the energy will be continuous over the length of the stream. If it is traveling slower though, then the water will be able to be held by boundaries along the length of the stream. The energy in this case will be discontinuous and will appear in "packets".

In this process of slowing down the implicit omnipresent-omnipotent-omniscient-omninurturing nature of light becomes or transforms into an explicit or emergent matter-past-present-future nature of light. This process of transformation creates a seed-equation that is present and synonymous with the Big Bang. The dynamics of a resulting universe are contained in the dynamics of such a seed-equation. The seed-equation will be explored in more detail in the section on mathematical foundations.

We have arisen in the field of light that has a finite speed. It is difficult therefore to feel the reality of the omnipresent-omnipotent-omniscient-omninurturing light. We are attuned to perceiving in the material world that has resulted due to the finite speed of light. But if we could step back into the fullness of light in its pure state at infinity from there we would likely see that there can be different universes created as a result of light selectively slowing down to different levels. The slow down to a different level will create a particular kind of universe.

Note that there have been experiments to slow down light so that it practically moves at a snail's pace. This slowing down has to be put into context. Even if light were to slow down to 1 mile per second, say, from 186,000 miles per second, keep in mind that it is possible that this change in speed is likely only a miniscule fraction of the change in speed from light traveling infinitely fast to light traveling at c, that is, 186,000 miles per second. So in effect what could be said is that when light is made to slow down in experiments, the range or band that it slows down to deviates only slightly from its slower speed in vacuum.

In this view the Big Bang is not an original creative event. It is likely a recurring creative event that allows something innate in light to express itself in time. But what is in light? What is in the omnipresent-omniscient-omnipotent-omninurturing nature of light? Everything we see around us, and an infinite more that we do not see.

Further, quanta being an accumulation of energy or possibility surfacing in the very process of light traveling at c, suggests the unique place it holds in a creation and in entering into anterior or contextual worlds of Light. This will be explored more fully subsequently.

Chapter 1.4: What is Light & What are Quanta?

When we think of light traveling at the speed of c, there are properties in the reality that emerge as a result of this. These properties as we have examined are of the nature of matter, the past, the present, and the future. These properties are the result of light traveling at c and so can be thought of as emerging from light.

But what do these properties actually mean? And further, how do these properties relate to the apparently quite different properties of omnipresence, omnipotent, omniscience, omninurturing that can appear in a reality where the speed of light is infinite?

Let us consider first the properties of light that emerge when light is traveling at c: the past, the present, the future, and matter.

What is the past? It is the perceivable result of all the work and effort that has taken place so far. It is the foundation upon which the present and future will be built. It represents a status quo, a stability, and even a rigidity, and given that it is the result of the long play of time, it will not easily be persuaded to become another thing. It can be thought of as that which the eye can see when it looks around it. There is "physicality" to what the eye can see and so the essence of the past is a kind of physical-ness. So, ingrained in light, is this ability to project or create physical-ness.

What is the present? It is the tremendous play of forces of all kinds to express themselves here and now. There is "vitality" that is present in this play and often it is the most energetic or forceful of the forces that will win out, as opposed to the most insightful or thoughtful. All the tremendous possibility of the future is seeking for expression now and so this essential vitality can also be thought of as a projection or possibility implicit in light.

What is the future? It is the inevitability of what will manifest. The great thoughts, the great ideas, the purpose, the possibilities will sooner or later express themselves in what we call the future. And the essence of this is thoughtfulness or a curiosity or a purpose that we can summarize as an essential "mentality". So embedded in light is this ability to project mentality.

And what is matter? It is the myriad crystallization into apparent diversity of the one essential reality of light, to allow for a play between these different sides or possibilities in an increasingly harmonious interaction. So its essence is "harmony", and this too can be thought of as an essential property projected from or implicit in light.

But what about when light is considered to move at an infinite speed? Then the properties that become apparent, as already explored, are those of omnipresence, omnipotence, omniscience, and omninurturing. But omnipresence has physical-ness to it, omnipotence has vitality to it, omniscience a mentality to it, and omninurturing a harmony to it. It may be said that physical-ness comes from omnipresence, vitality comes from omnipotence, mentality comes from omniscience, and matter comes from omninurturing.

So whether light is traveling at an infinite speed or the speed we know as c, there is something about the properties it projects, that in essence is the same. So let us refer to these essential properties made apparent through the worlds that are created, as Presence (from omnipresence), Power (from omnipotence), Knowledge (from omniscience), and Harmony (from omninurturing).

Further, quanta can be thought of as existing on that very border of the world or reality created due to Light traveling at c, and realities created as Light travels faster, which at its limit is an infinite speed. So quanta are a doorway or an interface into worlds of Light, and a doorway by which possibilities in deeper worlds of Light can express themselves here in this material realm.

Chapter 1.5: Structured Time, Structured Space, and Their Relation To Quanta

In considering the worlds that are created due to the way light moves or the play of light, we see properties that are projected because of it. In the finite world, the world that results from light traveling at the speed c, there is, relatively, smallness we can grasp and on which we can build. In the infinite world that results from light traveling at an infinite speed there is a vastness and fullness that is difficult to grasp.

The notions of time and space are something entirely different in both worlds, and it could be said that Space allows the full play of everything meant by Power, Knowledge, Harmony, and Presence to be seeded in it, and that Time allows that seeding to flower into fuller forms with its passage.

In other words both space and time are not just abstract concepts but are essentially highly structured to allow physical-ness, vitality, mentality, and matter to become Presence, Power, Knowledge, and Harmony.

Space allows all the possibilities present in Presence, Power, Knowledge, and Harmony and seeded in vast diversity, to evolve into more fullness through the time stages of physical-ness, vitality, mentality, and evolving matter.

Such a creation of space and time is synonymous with the creation of quanta. Quanta become the means for the possibilities inherent in the anterior worlds of Presence, Power, Knowledge, and Harmony to express themselves in a structured space and time. Quanta are therefore a passage into deeper worlds of Light, and a means for possibilities in these deeper realms to express themselves materially.

But further, in this view it may also be proposed that space and time being so structured by the four properties of Light, need also to be quantized. That is, space and time must be experienced as quanta as well.

Chapter 1.6: The Quantization of Space, Time, Matter, and Gravity

Chapter 1.1 proposed that quanta is the result of the slowing down of light, and as such becomes the basis for material expression. In other words, for matter to express itself requires quanta. But further, in Chapter 1.5 it was also proposed that space, consisting of seeds of the properties of light, would also require quanta to express itself. Philosophically this has to be in a world where light is traveling at a fraction of its possible speed. Further, it was also proposed that time is highly structured, expressing the growth of the seeds in space through definite phases. As such, growth through such phases can also be thought of as happening due to quantization that so allows phases to express themselves.

But if we take a deeper look at space, time, and matter in light of the properties of Light, it can be deduced that space, consisting of a vast array of seeds derived from the properties of Light is itself an expression of Light's property of Knowledge.

Time, bringing forth the meaning contained in the seeds, regardless of circumstance, and even being opposed by circumstance, can be thought of as Light's property of Power.

Matter itself, being a container in which space and time can allow deeper properties of Light to become materially tangible, must be an expression of Light's property of Presence.

But it is also known from Einstein's General Theory of Relativity that gravity is associated with mass and space, in that it is none other than a mass's instruction telling space how to curve, and again is nothing else that space's instruction telling mass how to move through it.

As such, where mass and space exist, there gravity has to exist as well. Hence it may be inferred that gravity is none other than an expression of Light's property of Harmony, which fixes the collective relationship between object and object.

But if as proposed matter, space, and time all need to be quantized in order to express themselves, then this must also be true of gravity.

The subsequent section on Mathematical Foundations will begin to explore this in greater detail.

Chapter 1.7: Functional-Richness, Super-Matter & Expanding Space

Quantization of space, time, energy, and gravity can be thought of as a means by which Light's properties of Knowledge, Power, Presence, and Harmony can manifest materially to so change the very fabric of experienced reality. Quantization happens because light travels at a fraction of the speed with which it can potentially travel. In the process there is a build-up of some aspect of light, depending on the quantization occurring.

The very bases of matter starting from fields, through quantum particles, through atoms and subsequently molecules can all ne modeled as emergences of the four-fold properties of Light. In the process of such emergence it would be noticed that each subsequent layer of material complexity contains or embodies greater functional-richness.

Hence it could be proposed that not only is the process of material and subsequent creation driven by over-arching meta-functions, but that the very variance in atoms, and molecules, and subsequently in living cells, for example is also driven by an increasing need to express functional-richness.

If Light contains within it all possibility, then it must be that a creation due to Light exists to fundamentally express all the possibilities contained within it in tangible form.

Space, is proposed as being a seeding ground for infinite variety to emerge. Hence it may also be proposed that as the process of creation is successful so that more and more properties inherent in Light actually get materialized, Space has an incentive to continue to expand to allow more seeds to be 'sowed'. Quantization of space then becomes the means by which that which is inherent in Light seeds itself. As already discussed time-quantization, energy-quantization, and gravity-quantization work with space-quantization to allow such Light-possibilities to manifest.

Since matter is the focal-point of the process of functional-complexification it is conceivable that the trajectory of matter and life experienced so far yields further to a "super-matter" which is just a foundation for an incredible and perhaps unending materialization of functional-richness. In such a view this functional-richness is the possible fate of this therefore expanding universe, so long as certain conditions are fulfilled. Through the use of a function-based mathematics this book is going to explore the conditions that will likely lead to establishing such a fate.

SECTION 2: THE MATHEMATICAL FOUNDATION

This section explores the mathematics required to interpret the universe as a play of functional-richness. The Light-Matrix describes the overarching structure from a ubiquitous unit to the vast diversity of life. Each layer in the matrix is then explored in more detail through the unit-point, the architectural forces, uniqueness of organizations and its emergence, and the summary inherent dynamics always available. The dynamics of possibility is explored through qualified determinism. The foundational Light-Space-Time Emergence equation is then explored which is leveraged in subsequent quantization analyses. Basic interpretations of Schrodinger's wave equation and Heisenberg's uncertainty principle are also explored.

This section will hence, specifically explore the following foundational mathematics:

- Light-Matrix – The Seed-Equation
- Nature of Point-Instant in a System
- Architectural Forces
- Uniqueness of Organizations
- Emergence of Uniqueness
- Inherent Dynamics of Any System
- Qualified Determinism
- Equation for Light-Space-Time Emergence
- Speed of Light and Quanta
- Interpreting Schrodinger's Equation
- Interpreting Heisenberg's Uncertainty Principle
- A Deeper Look at Quantization of Space, Time, Matter, and Gravity

Chapter 2.1: Light-Matrix – The Seed-Equation

This chapter will explore the Light-Matrix, a mathematical expression of the codification in Light formed at the time of the Big Bang. This Light-Matrix can be thought of as a seed-equation that provides insight into dynamics of the universe. The Light-Matrix codifies dynamics that may be experienced at the quantum border of realities created through Light traveling at different speeds, and further in the worlds that the quantum veil or window provide access to.

Nature of Light at U

As laid out in Section 1, it is perhaps fair to say that the speed of light has significant implications on the experienced nature of reality. The finiteness, c, at 186,000 miles per second in a vacuum (Perkowitz, 2011), creates an upper bound to the speed with which any object may travel also implying that objective reality will be experienced as a past, a present, a future from the point of view of that object (Einstein, 1995). These characteristics – a past, a present, a future – are implicit in the nature of light and become part of objective reality because of the speed of light. So the way we experience time seems definitely to be determined by c.

Further, c also creates a lower bound when inverted (1/c) being proportional, or arguably even determining Planck's constant, h, that pegs the minimum amount of energy or quanta required for expression at the sub-atomic level. Planck's constant, h, pegging the amount of energy required for expression, therefore may allow matter to form (Lorentz, 1925) and for the reality of nature with a past, present, and future, to also be experienced as a phenomena of connection between seemingly independent islands of matter. This characteristic of 'connection' is therefore also proposed to be implicit in the nature of light and becomes part of objective reality because of the speed of light.

But as explored in Chapter 1.4, a 'past' can also be viewed as established reality as defined by what the eye or other lenses of perception can see. It is the perceivable result of all the work and effort that has taken place so far. It is the foundation upon which the present and future will be built. It represents a status quo, a stability, and even a rigidity, and given that it is the result of the long play of time, it will not easily be persuaded to become another thing. It can be thought of as that which the eye can see when it looks around it. There is "physicality" to what the eye can see and so the essence of the past is a kind of physical-ness. So, ingrained in light, is this ability to project or create physical-ness.

Also as explored previously, the 'present' is the tremendous play of forces of all kinds to express themselves here and now. There is "vitality" that is present in this play and often it is the most energetic or forceful of the forces that will win out, as opposed to the most insightful or thoughtful. All the tremendous possibility of the future is seeking for expression now and so this essential vitality can also be thought of as a projection or possibility implicit in light.

The 'future' is the inevitability of what will manifest. The great thoughts, the great ideas, the purpose, the possibilities will sooner or later express themselves in what we call the future. And the essence of this is thoughtfulness or a curiosity or a purpose that we can summarize as an essential "mentality". So embedded in light is this ability to project mentality.

These implicit characteristics of the nature of light as experienced at the layer of reality so set up by a finite speed of light may hence be summarized by Equation 2.1.1, where c_U refers to the speed of light of 186,000 miles per second, that has created the perceived nature of reality, U, as expressed in Equation 2.1.1:

$$c_U: [Physical, Vital, Mental, Connection]$$

Eq. 2.1.1: Implicit Characteristics of the Nature of Light at U

Nature of Light at ∞

Exploring further, it is known however that at quantum levels the nature of reality is characterized by wave-particle duality. Light itself (Feynman, 1985) and matter (De Broglie, 1929) may be experienced as both particles and waves. But for matter to be experienced as waves implies that 'h' must have become a fraction of itself, $h_{fraction}$, to allow the concentration or possibility of quanta to have dispersed into wave-form. This further implies that c must have become greater than itself, c_N, such that the inequality specified by Equation 2.1.2 holds:

$$c_N > c_U$$

Eq. 2.1.2: Layer N, Layer U Inequality of Speed of Light

Note that what is implied here is that just as there is a nature of reality specified by U that is the result of the speed of light being 186,000 miles per second, so too there is another nature of reality specified by N that is the result of a speed of light greater than 186,000 miles per second.

This is akin to recent developments in physics with the notion of property spaces being separate from but influencing physical space as explored by Nobel Physicist Frank Wilczek (Wilczek, 2016). But further in "Slow Light" Perkowitz's recent treatment of today's breakthroughs in the science of light (Perkowitz, 2011) he states: "Although relativity implies that it's impossible to accelerate an object to the speed of light, the theory may not disallow particles already moving at speed c or greater. In the 1960's, Olexa-Myron P. Bilaniuk of Swarthmore College and E.C. George Sudarshan at Syracuse University began considering how to fit what they called "metaparticles" with speeds greater than c into the relativistic scheme. The approach was extended in 1967 by Gerald Feinberg (Feinberg, 1970), or Rockefeller and Columbia Universities, in his theoretical paper "Possibility of Faster-Than-Light-Particles," Feinberg also introduced the wonderful name "tachyons" for these hypothetical particles, from the Greek word "tachys" meaning swift." Perkowitz goes on to say how a flurry of papers have continued to appear about tachyons.

So light traveling at c_N may be possible. Current instrumentation, experience, and normal modes of thinking though, having developed as a bi-product of the characteristics so created in the layer of reality U may be inadequate to access N without appropriate modification. The notion of wave-particle duality already challenges the notion of normal thinking perhaps because wave-like phenomena could be a function of faster than c motion, and particle-like phenomena a function of less than or equal to c motion. That these may be happening simultaneously is reinforced by principles such as complementarity in which experimental observation may allow measurement of one or another but not of both (Whitaker, 2006).

But then taking this trend of a possible increase in the speed of light to its limit, this will result in a speed of light of infinite miles per second. The question is, what is the nature of reality when light is traveling at infinite miles per second? As explored in Section 1, in any space-time continuum light originating at any point will instantaneously have arrived at every other point. Hence light will have a full and immediate *presence* in that space-time continuum. Further, that light will *know* everything that is happening in that space-time completely instantaneously – that is know what is emerging, what is changing, what is diminishing, what may be connected to what, and so on - or have a quality of knowledge. It will connect every object in that space-time completely and therefore have a quality of connection or *harmony*. Finally nothing will be able to resist it or set up a separate reality that excludes it and hence it will have a quality of *power*.

These implicit characteristics of the nature of light as experienced at the layer of reality so set up by an infinite speed of light may hence be summarized by Equation 2.1.3, where c_∞ refers to the speed of light of ∞ miles per second, that has created the perceived nature of reality, ∞:

$$c_\infty : [Presence, Power, Knowledge, Harmony]$$

Eq. 2.1.3: Implicit Characteristics of Nature of Light at ∞ Speed

Transformation from ∞ to U

But by (2.1.3) it can also be noticed that 'physical' is related to Presence, 'vital' is related to Power, 'mental' is related to Knowledge, and 'connection' is related to Harmony.

The question then, is how do these apparent qualities at ∞ precipitate or become the physical-vital-mental-connection based diversity experienced at U? This may be achieved through the intervention or action of a couple of mathematical transformations acting on the implicit characteristics of nature of light at ∞ speed as summarized by (2.1.3).

First, the essential characteristics of Presence, Power, Knowledge, Harmony that it is posited exist at every point-instant by virtue of the ubiquity of light at ∞ will need to be expressed as sets with up to infinite elements. Second, elements in these sets will need to combine together in potentially infinite ways to create a myriad of seeds or signatures that then become the source of the immense diversity experienced at U. This suggests that all that is seen and experienced at U may be nothing other than 'information' or 'content' of light and as such that there are fundamental mathematical symmetries at play where everything at U is essentially the same thing that exists at ∞.

Assuming that the first transformation occurs at a layer of reality K where the speed of light is c_K, such that $c_U < c_K < c_\infty$, this may be expressed by Equation 2.1.4:

$$c_K: [S_{Pr}, S_{Po}, S_K, S_H]$$

Eq. 2.1.4: The First Transformation at Layer K

S_{Pr} signifies 'Set of Presence', S_{Po} signifies 'Set of Power', S_K signifies 'Set of Knowledge', S_H signifies 'Set of Harmony/Nurturing'.

Assuming that the second transformation occurs at a layer of reality N where the speed of light is c_N, such that $c_U < c_N < c_K < c_\infty$, this may be expressed by Equation 2.1.5:

$c_N\colon f(S_{Pr}\ x\ S_{Po}\ x\ S_K\ x\ S_H)$

Eq. 2.1.5: The Second Transformation at Layer N

The unique seeds are therefore a function, f, of some unique combination of the elements in the four sets S_{Pr}, S_{Po}, S_K, S_H.

The relationship between the layers of light may be modeled by the following matrix in Equation 2.1.6:

$$Light_{Matrix} = \begin{vmatrix} c_\infty\colon [Pr, Po, K, H] \\ (\downarrow\ R_{C_K} = f(R_{C_\infty})) \\ c_K\colon [S_{Pr}, S_{Po}, S_K, S_H] \\ (\downarrow\ R_{C_N} = f(R_{C_K})) \\ c_N\colon f(S_{Pr}\ x\ S_{Po}\ x\ S_K\ x\ S_H) \\ (\downarrow\ R_{C_U} = f(R_{C_N})) \\ c_U\colon [P, V, M, C] \end{vmatrix}$$

Eq. 2.1.6: Light-Matrix

The matrix should be read from the top row down to the bottom row as indicated by the ↓ between rows, and suggests a series of transformations leading from the ubiquitous nature of light implicit in a point – presence, power, knowledge, harmony - to the seeming diversity of matter observed at the layer of reality U which is fundamentally the same presence, power, knowledge, and harmony projected into another form of itself.

The first transformation is summarized by Equation 2.1.7:

$$R_{C_K} = f(R_{C_\infty})$$

Eq. 2.1.7: Light-Matrix First Transformation

This is suggesting that the reality at the layer specified by the speed of light c_K, R_{C_K} is a function of the reality at the layer specified by the speed of light c_∞. This transformation translates the essential nature of a point into the sets described in Equation 2.1.4.

The second transformation is summarized by Equation 2.1.8:

$$R_{C_N} = f(R_{C_K})$$

Eq. 2.1.8: Light-Matrix Second Transformation

This is suggesting that the reality at the layer specified by the speed of light c_N, R_{C_N} is a function of the reality at the layer specified by the speed of light c_K. This transformation combines elements of the sets into unique seeds as suggested by Equation 2.1.5.

The third transformation is summarized by Equation 2.1.9:

$$R_{C_U} = f(R_{C_N})$$

Eq. 2.1.9: Light-Matrix Third Transformation

This is suggesting that the reality at the layer specified by the speed of light c_U, R_{C_U} is a function of the reality at the layer specified by the speed of light c_N. This transformation builds on the unique seeds suggested by (2.1.5) to create the diversity of U as specified by (2.1.1).

In this framework the notion of wave-particle duality hence may become complementary block-field-wave-particle "quadrality" where block refers to phenomenon resident to ∞, field to phenomenon resident to N, wave to phenomenon resident to K, and particle to phenomenon resident to U.

The block is all the reality always present behind the surface and is captured by (2.1.3) or the top line in the Light-Matrix as expressed in (2.1.6). The field is captured by (2.1.4) or the second line from the top in (2.1.6) and can be thought of as layers of possibility existing in each of the sets. The wave is captured by the creation of seeds represented by (2.1.5) or by the third line in (2.1.6). The particle that is apparently disconnected from the whole is captured by (2.1.1) or the bottom line in (2.1.6).

So what we arrive at is a fundamental Light-Matrix that suggests key dynamics for different layers of Light. Quanta, being at the interface of Layer U and Layer N are so positioned as being subject to multiple sets of dynamics.

Chapter 2.2: Nature of a Point-Instant in a System

The 'point-instant' captures the inherent innovation that appears to exist in the system and is represented by the top-line in Equation (2.1.6), the Light-Matrix. This innovation is a function of the properties of light derived in Equation (2.1,3) - Presence, Power, Knowledge, and Harmony – and reproduced here for convenience:

$c_{\infty}: [Presence, Power, Knowledge, Harmony]$

The 'point' aspect of the point-instant suggests the space-dimension and that space is seeded with the possibilities inherent in the properties of Presence, Power, Knowledge, and Harmony. The 'instant' aspect suggests the time-dimension and gives insight into the process of emergence that the possibilities in space progressively surface as.

In its point-instant, Presence-Power-Knowledge-Harmony wholeness, Presence allows emergences to continue to develop as per the possibilities implicit in the past-present-future or physical-vital-mental pathway.

Beginning to translate this into an equation, the notation $System_{Pr}$ is given to system-presence. This system-presence is true across any considered Time-Space continuum starting from a time-space boundary '0' to a time-space boundary 'N'. This notion is characterized by the notation $TS_{0 \rightarrow N}$. Within that boundary from 0 to 'N', the 'presence' is such that it will always seize an opportunity to cause a shift from the physical-leading to the vital-leading, and from the vital-leading to the mental-leading. Research shows (Malik, 2009) that greater degrees of freedom is afforded by such a shift.

The notion that the presence seizes on opportunity as characterized by the notation:

$$\begin{matrix} Presence \\ \downarrow \\ Opportunity \end{matrix}$$

The shift from physical-leading to vital-leading and vital-leading to mental-leading is characterized by:

$$\begin{matrix} P_L & \rightarrow & V_L \\ V_L & \rightarrow & M_L \end{matrix}$$

Hence in this approach it is suggested that:

$$System_{Pr} \equiv TS_{0 \rightarrow N} \left[\begin{matrix} Presence \\ \downarrow \\ Opportunity \end{matrix} \begin{bmatrix} P_L & \rightarrow & V_L \\ V_L & \rightarrow & M_L \end{bmatrix} \right]$$

But there is something else about this Presence as well. All other developments take place in it. That is, it provides a container of sorts in which the plays of system-power, system-knowledge, and system-harmony/system-nurturing can take place. This notion is summarized by the notation:

$$Container_{\begin{bmatrix} System_P \\ System_K \\ System_N \end{bmatrix}}$$

Hence, combining these various components, an equation for 'system-presence', Equation 2.2.1, arises:

$$System_{Pr} \equiv TS_{0 \rightarrow N} \left[\begin{matrix} Presence \\ \downarrow \\ Opportunity \end{matrix} \begin{bmatrix} P_L & \rightarrow & V_L \\ V_L & \rightarrow & M_L \end{bmatrix} \ \& \ Container_{\begin{bmatrix} System_P \\ System_K \\ System_N \end{bmatrix}} \right]$$

Eq 2.2.1: System Presence

In its point-instant, Presence-Power-Knowledge-Harmony wholeness, Power allows emergences to continue to happen in spite of tremendous oppositions of all kinds; this too, regardless of field or area.

Constructing an equation for system-power, the notation $System_P$ is used to represent system-power. Any endeavor will always be met with resistances of various kinds. The resistances that arise along the physical dimension are referred to as P_R. The resistances that arise along the vital dimension are referred to as V_R. The resistances that arise along the mental dimension are referred to as M_R. In the fruition of any endeavor one or all of these types of resistances may arise. Further, resistance of one kind often feeds on resistance of another kind, and to generalize the resistances encountered in an endeavor may be characterized as the product of the three types of resistance

$$P_R \ * \ V_R \ * \ M_R$$

These resistances arise across any considered Time-Space boundary from 0 to 'N', and therefore it may be said that the power of the system is such that:

$$power > \sum_{TS=0}^{N} P_R \ * \ V_R \ * \ M_R$$

An equation for 'system-power', Equation 2.2.2, hence, is the following:

$$System_P \equiv power > \sum_{TS=0}^{N} P_R \ * \ V_R \ * \ M_R$$

Eq 2.2.2: System Power

In its point-instant, Presence-Power-Knowledge-Harmony wholeness, Knowledge orchestrates emergences to continue to happen by leveraging the right instruments and circumstances.

Translating this into an equation, the notation, $System_K$, is used for system-knowledge. This $System_K$ is such that it leverages the right instrumentation and circumstance to bring about the progress that is possible. This concept of 'instrumentation' is denoted by the subscript 'I'. The concept of 'circumstance' is denoted by the subscript 'C'. Both instrumentation and circumstance can be of a physical, vital, or mental type and this possibility is denoted by:

$$\begin{bmatrix} P_{I,C} \\ V_{I,C} \\ M_{I,C} \end{bmatrix}$$

Further, the notion that the 'knowledge' is such that it 'leverages' the right instrumentation and circumstance is depicted by:

$$\begin{matrix} Knowledge \\ \downarrow \\ Leverage \end{matrix}$$

This act of leveraging results in a fundamental shift so that the physical-leading yields to the vital-leading, and the vital-leading yields to the mental-leading. Hence:

$$\begin{matrix} Knowledge \\ \downarrow \\ Leverage \end{matrix} \begin{bmatrix} P_{I,C} \\ V_{I,C} \\ M_{I,C} \end{bmatrix} \rightarrow \begin{bmatrix} P_L & \rightarrow & V_L \\ V_L & \rightarrow & M_L \end{bmatrix}$$

Since this behavior may exist across any Time-Space continuum an equation for system-knowledge, Equation 2.2.3, is suggested:

$$System_K \equiv TS_{0\to N}\left[\begin{matrix} Knowledge \\ \downarrow \\ Leverage \end{matrix}\begin{bmatrix} P_{I,C} \\ V_{I,C} \\ M_{I,C} \end{bmatrix} \to \begin{bmatrix} P_L & \to & V_L \\ V_L & \to & M_L \end{bmatrix}\right]$$

Eq 2.2.3: System Knowledge

In its point-instant, Presence-Power-Knowledge-Harmony wholeness, Harmony or nurturing allows emergences to continue to happen with more and more degrees of freedom to come to the surface.

This characteristic of this implicit-nurturing may be referred to as 'system-nurturing'. Like the other characteristics it is suggested to exist across a Time-Space continuum. This is depicted by:

$$TS_{0\to N}$$

There is an action of nurturing such that any state is always advanced to a higher level. This is depicted by:

$$\coprod_{Nurturing}\begin{pmatrix} P_- & M_+ \\ V_- & V_+ \\ M_- & P_+ \end{pmatrix}$$

Hence, there is a 'union', depicted by 'U' that 'nurtures' the negatives towards their positives.

Further, there is an increasing action of nurturing such that the possibility of integration is always increased to form a larger and larger basis. This increasing basis is depicted as being modulated by the polar coordinates 'r' and 'θ', where r is the radius which increases from an initial value of '0', and 'θ' is an angle from '0' to '360'.

Hence, the equation of system-nurturing, Equation 2.2.4, is depicted as:

$$System_N \equiv TS_{0\rightarrow N}\left(\coprod_{Nurturing}\begin{pmatrix} P_- & M_+ \\ V_- & V_+ \\ M_- & P_+ \end{pmatrix} mod\,(r,\theta)\right)$$

Eq 2.2.4: System Nurturing

It is suggested that these four characteristics exist across any system, and to denote this it is generalized that every point in any system is embedded with this four-fold intelligence. It is suggested that this four-fold intelligence is resident in every instant-spot of the system. It is suggested that to be able to leverage or activate this four-fold intelligence at will is the ultimate act of innovation.

Such a depiction of models, as Bar-Yam states in "From Big Data to Important Information" (Bar-Yam, 2016), 'is "valid" only because of the irrelevance of details...If we want to say anything meaningful about a system—meaningful in the sense of scientific replicability or in terms of utility of knowledge—the only description that is important is one that has universality, that is, is independent of details. There is no utility to information that is only true in a particular instance. Thus, all of scientific inquiry should be understood as an inquiry into universality—the determination of the degree to which information is general or specific.'

Chapter 2.3: Architectural Forces

The characteristics embedded in a point as in (2.1.3) suggest a possibility that is hard to fathom. One can only glimpse the extraordinary nature embedded in a point. And yet it can be suggested that this extraordinary nature is barely visible unless the right analytical lens of the sort being suggested here is first set up. Further, it is suggested that this extraordinary nature is responsible for a broader set of architectural forces that exist behind the visible face of things.

Hence, system-presence, system-power, system-knowledge, and system-nurturing that define the nature of every point in our system, become more tangible as a broader set of architectural forces that emanate from each of them.

Considering system-presence, here is a characteristic that appears to be everywhere (Malik, 2009) at the service of all the constructs that develop within it. There is a diligence and perseverance by which any opportunity for progress is seized. Further, if one considers the extraordinary detail that appears in any construct, whether an atom, a body, a planet, or a galaxy, one is struck by the high degree of perfection that surfaces in this presence.

So if one contemplates the nature of this system-presence there is a set of forces that surface. Depicting such a set as $S_{System_{Pr}}$, one can arrive at elements such as Service, Perfection, Diligence, Perseverance, amongst others, that are part of this set. Hence, the set can be described by Equation 2.3.1:

$$S_{System_{Pr}} \ni [Service, Perfection, Diligence, Perseverance, \ldots]$$

Eq 2.3.1: Set of System Presence

Similarly, considering the characteristic of system-power, one can hypothesize that there is a family of forces that emanates from it. The kinds of forces may be thought of as Power, Courage, Adventure, Justice, amongst others. The set for system-power can hence be depicted by Equation 2.3.2:

$$S_{System_P} \ni [Power, Courage, Adventure, Justice, \ldots]$$

Eq 2.3.2: Set of System Power

Similarly, considering the system-knowledge as the root of various powers that emanate from it, one may characterize the set for system-knowledge by Equation 2.3.3:

$$S_{System_K} \ni [Wisdom, Law\ Making, Spread\ of\ Knowledge \ldots]$$

Eq 2.3.3: Set of System Knowledge

The set for system-nurturing is depicted by Equation 2.3.4:

$$S_{System_N} \ni [Love, Compassion, Harmony, Relationship\ \ldots]$$

Eq 2.3.4: Set of System Nurturing

Chapter 2.4 describes how unique signatures or seeds for any type of organization can be built by leveraging the elements in the sets of architectural forces.

Chapter 2.4: Uniqueness of Organizations

The hypothesis is that every organization, whether an atom, cell, person, team, corporation, market, or country is unique and that this uniqueness can be specified in terms of elements of the derived sets for power, knowledge, presence, and nurturing. This hypothesis for uniqueness stems from observations at multiple levels.

At the sub-atomic level Nobel Laureate Wolfgang Pauli's 'Pauli Exclusion Principle' states that no two similar fermions, which include fundamental particles with half-integer spin such as protons, neutrons, and electrons, can occupy the same quantum states simultaneously (Pauli, 1964). Spin has to do with the angle that the particle has to rotate through before being symmetrical with its original state. Half-integer spin particles need to rotate through 720 degrees before being symmetrical with their original state. The implication of the Pauli Exclusion Principle is that fundamental structure and consequently stability comes into being at the atomic level, which as is evident in the periodic table allows the separation of function related to form. This stability related to the underlying structure of atoms implies the basis of uniqueness and diversity. In the absence of the Exclusion Principle matter would just be a dense soup (Hawking, 1988) with particles occupying overlapping space.

At the observable level uniqueness is evident from the immense diversity of distinct species on earth (Mora, 2011) estimated to be over 2 million, and further the uniqueness of every member of each species. This member-level uniqueness is suggested by the difference in non-coding regions of the DNA that may vary in their sequence by about 1 to 4 percent, which in turn result in unique protein binding sequences of each human

(Snyder, 2010), as an example, which in turn results in unique observable qualities.

At the astronomical level Einstein's Special Theory of Relativity (Einstein, 1995) suggests that every coordinate system potentially has its own space-time rendering as opposed to there being one absolute space and time. This implies the notion of uniqueness as an implicit property of space.

The four properties explored in the Chapter 2.1 and 2.2 define the source of that innovation. From this source emanate 4 sets of forces that suggest the boundaries of that innovation.

Assuming then that the fount of uniqueness is system-presence, a general equation for organizations that belong to the family of system-presence can be derived. Such uniqueness can be depicted as Sig_x where the subscript 'x' refers to the source family, and 'Sig' or signature to 'uniqueness'. Hence the uniqueness of an organization in the family of system-presence would be notated by $Sig_{system_{Pr}}$.

In line with the development of properties of a point and the precipitating architectural forces as discussed in Chapter 2.1 and 2.2 respectively, an approach to constructing such uniqueness is to assume a primary factor X that drives the uniqueness that belongs to the set $S_{system_{Pr}}$. Further, assume that the uniqueness is qualified by a number of secondary factors Y that may belong to any of the 4 sets - $S_{system_{Pr}}, S_{system_{P}}, S_{system_{K}}, S_{system_{N}}$. The primary factor X would have a greater weightage than any of the secondary factors Y. The weightage of X hence could be depicted by the number 'a', and the weightage of Y a number $'b_{o-n}'$, such that a > b. Further, the secondary element can repeat from '0 – n' times, and is hence depicted as $\overline{Yb_{0-n}}$.

The equation, Equation 2.4.1, hence for a unique organization derived from the family of system-presence is:

$$Sig_P = Xa + \overline{Yb_{0-n}} \quad where \begin{bmatrix} X \in [S_{System_{Pr}}] \\ Y \in [S_{System_{Pr}}, S_{System_P}, S_{System_K}, S_{System_N}] \\ a, b \; are \; integers; a > b \end{bmatrix}$$

Eq 2.4.1: System Presence Based Unique Organization

Similarly, an equation, Equation 2.4.2, for a unique organization derived from the family of system-power is:

$$Sig_V = Xa + \overline{Yb_{0-n}} \quad where \begin{bmatrix} X \in [S_{System_P}] \\ Y \in [S_{System_{Pr}}, S_{System_P}, S_{System_K}, S_{System_N}] \\ a, b \; are \; integers; a > b \end{bmatrix}$$

Eq 2.4.2: System Power Based Unique Organization

An equation, Equation 2.4.3, for a unique organization derived from the family of system-knowledge is:

$$Sig_M = Xa + \overline{Yb_{0-n}} \quad where \begin{bmatrix} X \in [S_{System_K}] \\ Y \in [S_{System_{Pr}}, S_{System_P}, S_{System_K}, S_{System_N}] \\ a, b \; are \; integers; a > b \end{bmatrix}$$

Eq 2.4.3: System Knowledge Based Unique Organization

An equation, Equation 2.4.4, for a unique organization derived from the family of system-nurturing is:

$$Sig_I = \overline{Xa + Yb_{0-n}} \quad where \begin{bmatrix} X \in [S_{System_N}] \\ Y \in [S_{System_{Pr}}, S_{System_P}, S_{System_K}, S_{System_N}] \\ a, b \ are \ integers; a > b \end{bmatrix}$$

Eq 2.4.4: System Nurturing Based Unique Organization

The four preceding equations can be generalized by Equation 2.4.5:

$$Sig = \overline{Xa + Yb_{0-n}} \quad where \begin{bmatrix} X \in [S_{System_{Pr}}, S_{System_P}, S_{System_K}, S_{System_N}] \\ Y \in [S_{System_{Pr}}, S_{System_P}, S_{System_K}, S_{System_N}] \\ a, b \ are \ integers; a > b \end{bmatrix}$$

Eq 2.4.5: Generalized Equation for Unique Organization

Having considered the structure of uniqueness, the next question is how does such uniqueness emerge? This is discussed in Chapter 2.5.

Chapter 2.5: Emergence of Uniqueness

While the uniqueness of organizations as represented by the Signature is a seed, like any seed there is a process for its emergence (Kaufmann, 1995; Portugali, 2012; Yates, 2012), and the uniqueness will often be hidden or very much behind the scene until certain conditions are fulfilled (Malik, 2009).

The implicit nature of Time and Space suggest a universal developmental model that provides a cue as to the process for emergence (Deep Order Mathematics Videos, 2016). In this model the four sets of architectural forces already described form a pool in space, as it were, from which possibility arises. Possibility itself is unique from point to point and is governed by the Equation for Uniqueness (Equations 2.4.1 through 2.4.5) described in the previous chapter.

Hence it is observed that initially the uniqueness takes a 'physical' form, moving on to a 'vital' form, and then onto a 'mental' form. Once the orientations implicit in each of these phases are assimilated, then the uniqueness takes on an 'integral' form. The integral form is a threshold phase, and allows the uniqueness suggested by the Signature to emerge in fuller force or in its 'force' form. The final phase is the 'contextual form' that allows the signature to act with impunity within a considered context.

Mathematically, if an organization exists at the physical phase, it may be suggested that its signature or uniqueness is modulated by the constant 'π'. π is the seed of a circle or sphere and can be thought of as defining behavior that is tightly bound. Within such a tightly bound volume it will likely not even be apparent what the uniqueness of an organization necessarily is. Assuming the uniqueness to be defined by the derived question *Sig*, the physical-level (P) behavior can be described by the following equation-segment where 'mod' signifies modulated-by:

$P: Sig * mod\ (\pi)$

If an organization exists at the vital level, it may be suggested that its uniqueness is modulated by the Euler-constant 'e'. e is at the root of exponential behavior. The vital by definition is about assertive and aggressive growth the symbol of which is 'e'. Hence vital-level (V) modulation (represented by 'mod') can be described by the following equation-segment:

$V: Sig * mod\ (e)$

If an organization exists at the mental level, it may be suggested that its uniqueness is modulated by the Gaussian Distribution 'G'. G summarizes rational behavior with a key direction followed by most, and directions more on the edge followed by outliers. Mental-level dynamics are arguably quite similar, and it can be suggested are best modeled by such a distribution (Salkind, 2007). Mental-level (M) modulation (mod) can hence be described by the following equation-segment:

$M: Sig * mod\ (G)$

The physical, the vital, and the mental levels are orientations in which patterns of perceiving, being, behaving are set in their ways. Each pattern has its purpose and its limitation and it can be argued that being able to learn from each orientation and yet being able to move beyond that, is the next logical step in any developmental model. The integral level hence, is about being able to leverage each of the patterns that naturally arise at the three preceding levels at will, and about further, being able to integrate these and arrive at new ways of perceiving and being.

Mathematically such behavior may be represented as being an integrative function ($\int x$) where 'x' is the ability to move between the patterns emanating from G, e, π, at will, represented by $\overline{G, e, \pi}$. Integral-level (I) modulation (mod) of uniqueness (*Sig*) can hence be represented by the following equation-segment:

$$I:\ Sig * mod\ (\int \overline{G, e, \pi}\ )$$

The condition of overcoming any fixed and limiting patterns is the prerequisite for the emergence of 'Force' or for entering into the force-level. At this level the uniqueness behind the particular development being considered can emerge in its purity and become a truly creative dynamic. This aspect of creativity that is in a sense not bound by circumstance may be represented by the constant 'c', the speed of light in a vacuum (Perkowitz, 2011) which is an upper limit of the layer that systems practically operate in. Force-level (F) modulation (mod) of uniqueness (*Sig*) can hence be represented by the following equation-segment:

$$F:\ Sig\ * mod\ (c)$$

Once the signature of an organization arises and continues to exercise itself in its purity, it achieves contextual-mastery (C) and is able to exercise itself as though the context it is acting in, that can vary in scale and complexity, were all of the same substance as itself. This equality may be represented by the integrative function '$\int = 1$'. The equation-segment that notates this contextual-level (C) modulation (mod) applied to organizational uniqueness (*Sig*) is hence:

$$C:\ Sig * mod\ (\int\ = 1)$$

Piecing all the equation-segments together the equation for the emergence of uniqueness (Sig_E), where 'X' can be any of the discussed modulations at the respective development-model levels (P, V, M, I, F, C), is hence summarized by Equation 2.5.1:

$$Sig_E = X \begin{vmatrix} C: Sig * mod\ (\int = 1) \\ F: Sig\ mod\ (c) \\ I: Sig\ mod\ (\int \overline{G, e, \pi}\ ) \\ M: Sig * mod\ (G) \\ V: Sig * mod\ (e) \\ P: Sig * mod\ (\pi) \end{vmatrix}$$

Eq 2.5.1: Emergence of Uniqueness

The power of virtual worlds in engaging people and promoting learning as described in "Why Virtual World Matter" (Thomas & Brown, 2009) can be seen as parallel to the emergence of uniqueness as described by Equation 2.5.1. As described by the authors there is a process of selectively leaving behind part of oneself (movement through the P, V, M levels) to recreate one's identity (movement through level I), and then engaging with others in a shared discourse and culture (movement through levels F, C), which becomes very meaningful for people.

Chapter 2.6: Inherent Dynamics of Any System

So far the inherent innovation that exists at the system level and summarized by the nature of a point has been considered. Further, how this deep fount of innovation is present everywhere and how sets that make more practical the range of creative forces available in each of the four components of a point have also been considered. These architectural forces further define the possibility inherent in any system. Leveraging these sets of forces an equation for the uniqueness of an organization, regardless of scale, was arrived at.

In some sense the precipitation of innovation from the barely perceptible nature of the ubiquitous point, to how this reveals a play of forces, to how organizations take their seed and grow from these forces, has been traced.

It is useful to now turn full-circle to return to the initial orientations that allowed so much to be suggested about the nature of innovation in the first place. It is useful to look deeper into the nature of the physical, the vital, the mental, and the integral, and to derive equations that in effect will provide further insight into the dynamics of innovation inherent in these orientations.

Hence, starting with the physical, an equation, Equation 2.6.1, is summarized as:

$$Physical = \begin{bmatrix} M_3 \rightarrow System_{Pr} \\ (\uparrow F \rightarrow I) \\ M_2 \rightarrow S_{Systemi_{Pr}} \\ (\uparrow Sig \rightarrow F) \\ M_1 \rightarrow Sig_P \\ (\uparrow > P_{P)} \\ U \rightarrow Physical_U \end{bmatrix} TC \rightarrow Physical_T$$

$$Where \begin{bmatrix} Physical_U \ni [inertia, \square ethargy, status\ quo, \dots] \\ Physical_T \ni [adaptability, durability, strength, \dots] \end{bmatrix}$$

Eq 2.6.1: Inherent Dynamism in Physical

Essentially this equation is laying out the conditions of moving from the untransformed or negative physical state represented by $Physical_U$ to the transformed or positive physical state represented by $Physical_T$.

The first matrix should be read from the bottom to the top:

$$\begin{bmatrix} M_3 \rightarrow System_{Pr} \\ (\uparrow F \rightarrow I) \\ M_2 \rightarrow S_{System_{Pr}} \\ (\uparrow Sig \rightarrow F) \\ M_1 \rightarrow Sig_P \\ (\uparrow > P_{P)} \\ U \rightarrow Physical_U \end{bmatrix}$$

Hence, at the bottom is the starting point '$U \rightarrow Physical_U$' which identifies the default or untransformed (U) level of the physical. The next row up, $(\uparrow > P_P)$, states that when the patterns of the untransformed physical (P_P) have been overcome (>), movement to the next level (↑) is facilitated. Breaking through to the next level, $M_1 \rightarrow Sig_P$, allows its dynamics to become active. Hence, the signature or uniqueness of the physical (Sig_P) becomes active at meta-level 1 (M_1). As this signature becomes more like a Force ($Sig \rightarrow F$), the conditions for breakthrough (↑) to the next level are achieved. This next level is referred to as meta-level 2 (M_2), and indicates that the architectural forces represented by the set of system-presence ($S_{System_{Pr}}$) have become more consciously active. When this Force becomes Integral ($F \rightarrow I$) then the conditions for breakthrough (↑) to the next level are achieved. The next level is notated as M_3 for meta-level 3, and the dynamics here indicate that the equation for system-presence becomes active. Becoming active basically means that the respective meta-level dynamic begins to act at the once 'untransformed' level (U) further modifying it. Modification or transformation began when M_1 became active. Transformation is accelerated when M_2 becomes active, and even further accelerated when M_3 becomes active. Note that the matrix is essentially semiotic in that adjacent levels exist in paired relationships and inter-related functionality (Usó-Doménech et al, 2016).

The notion of meta-layers is being explored by contemporary physicists and Erwin Laszlo in his book Self-Actualizing Cosmos (Laszlo, 2014) summarizes some of these developments: "Physicists describe the domain that underlies and embeds the particles, fields, and forces of the universe variously as quantum vacuum, physical spacetime, nuether, zero-point field, grand-unified field, cosmic plenum, or string-net liquid." Note that 'nuether' refers to a sub-quantum level of reality (Pearson, 1997). Laszlo goes on to describe a revolutionary discovery by Nika Arkani-Hamed of Princeton's Institute of Advanced Study, of a geometrical object, the amplituhedron (Arkani-Hamid et al, 2012), which is not in space-time but governs space-time so that it "appears that spatio-temporal phenomena are the consequence of geometrical relationships in a deeper dimension of physical reality". A deeper dimension of a physical layer suggests synonymity with a meta-layer.

The rate of the transformation can be better envisioned when considering action of the Transformation Circle, or TC. The TC can be thought of as 4 concentric circles, with M_3 at the center. M_3 is surrounded by M_2, which is surrounded by M_1. The outer circle is U. If TC is considered to be a clock, than at time 't = 0', the physical' can be thought of as being entirely in U. The clock starts ticking only when some initial patterns P_P are overcome $(> P_P)$. From this point on as time proceeds the conditions for breakthrough become riper, and a sinusoidal wave begins to integrate more of the concentric circles together. The sinusoid wave (sin) is itself modulated by an euler function, e^x, where 'x' is determined by the strength to overcome patterns (↑) which will likely vary over time but will likely tend to be positive once the clock has started ticking because of the joy experienced with progressive movement. Being that the limit is the outer boundary of the concentric circles, there is further modulation by π until the 4 concentric circles have been integrated. TC, hence, may be represented by Equation 2.6.2:

$$TC \equiv (> P_P) \rightarrow \text{mod}\,(\text{sin}, e^x, \pi)$$

Eq 2.6.2: Transformation Circle

Hence, the initial nature of the physical that may be characterized by the set comprising of elements such as, lethargy, acceptance of the status quo, amongst other such elements ($Physical_U \ni [inertia, lethargy, status\ quo, \ldots]$), transforms into a physical more characterized by elements such as adaptability, durability, strength, and so on ($Physical_T \ni [adaptability, durability, strength, \ldots$). This transformation represents the inherent innovation-dynamic within the Physical.

Similarly, the equation for the 'Vital', Equation 2.6.3, also shows the built-in transformation that represents the innovation-dynamic within the vital:

$$Vital = \begin{bmatrix} M_3 \rightarrow System_P \\ (\uparrow F \rightarrow I) \\ M_2 \rightarrow S_{System_P} \\ (\uparrow Sig \rightarrow F) \\ M_1 \rightarrow Sig_V \\ (\uparrow > P_{V)} \\ U \rightarrow Vital_U \end{bmatrix} TC \rightarrow Vital_T\ ,$$

$$Where \begin{bmatrix} Vital_U \ni [aggression, self\ centeredness, exploitation, \dots] \\ Vital_T \ni [energy, support, adventure, enthusiasm, \dots] \end{bmatrix}$$

Eq 2.6.3: Inherent Dynamism of Vital

The equation for the 'Mental', Equation 2.6.4, is similarly summarized as:

$$Mental = \begin{bmatrix} M_3 \rightarrow System_S \\ (\uparrow F \rightarrow I) \\ M_2 \rightarrow S_{System_S} \\ (\uparrow Sig \rightarrow F) \\ M_1 \rightarrow Sig_M \\ (\uparrow > P_{M)} \\ U \rightarrow Mental_U \end{bmatrix} TC \rightarrow Mental_T$$

$$Where \begin{bmatrix} Mental_U \ni [fixation, fundamentalism, fragmentation, \dots] \\ Mental_T \ni [understanding, imagination, inspiration, \dots] \end{bmatrix}$$

Eq 2.6.4: Inherent Dynamism of Mental

The equation for the 'Integral', Equation 2.6.5, is similarly summarized as:

$$Integral = \begin{bmatrix} M_3 \rightarrow System_N \\ (\uparrow F \rightarrow I) \\ M_2 \rightarrow S_{System_N} \\ (\uparrow Sig \rightarrow F) \\ M_1 \rightarrow Sig_I \\ (\uparrow > P_{I)} \\ U \rightarrow Integral_U \end{bmatrix} TC \rightarrow Integral_T$$

$$Where \begin{bmatrix} Integral_U \ni [possession, usurpation, hidden\ agendas, \ldots] \\ Integral_T \ni [appreciation, shift\ POV, MPV, synthesis, \ldots] \end{bmatrix}$$

Eq 2.6.5: Inherent Dynamism of Integral

The preceding equations can be generalized by Equation 2.6.6:

$$Innovation_{orientation-x} = \begin{bmatrix} M_3 \rightarrow System_X \\ (\uparrow F \rightarrow I) \\ M_2 \rightarrow S_{System_X} \\ (\uparrow Sig \rightarrow F) \\ M_1 \rightarrow Sig_x \\ (\uparrow > P_{x)} \\ U \rightarrow x_U \end{bmatrix} TC \rightarrow x_T, where \begin{bmatrix} x_U \ni [\ldots] \\ x_T \ni [\ldots] \end{bmatrix}$$

Eq 2.6.6: Generalized Equation of Innovation

In this generalized equation, $Innovation_{orientation-x}$, refers to the inherent innovation within a specific orientation. Orientation refers to the physical, the vital, the mental, or the integral.

Further, the notion of a core-matrix can be summarized by the following equation, Equation 2.6.7:

$$Core\_matrix = \begin{bmatrix} M_3 \rightarrow System_X \\ (\uparrow F \rightarrow I) \\ M_2 \rightarrow S_{System_X} \\ (\uparrow Sig \rightarrow F) \\ M_1 \rightarrow Sig_x \\ (\uparrow > P_{x)} \\ U \rightarrow x_U \end{bmatrix}$$

Eq 2.6.7: Core Matrix

One of the corollaries of the Generalized Equation of Innovation is that if the source of innovation is more influenced by a meta-level there will be simultaneity of innovation or emergence that becomes apparent at U. The higher the meta-level, the more likely that this simultaneity will be wider spread. In his book Where Good Ideas Come From: The Natural History of Innovation (Johnson, 2010) Johnson states: "A brilliant idea occurs to a scientist or inventor somewhere in the world, and he goes public with his remarkable finding, only to discover that three other minds had independently come up with the same idea in the past year." He refers to an essay "Are Inventions Inevitable" (Ogburn & Thomas, 1922) which uncovered 148 instances of similar yet independent innovation, most of them occurring within the same decade. Some examples include sunspots that were simultaneously discovered in 1611 by four scientists in four different countries, and the law of conservation of energy that was formulated separately four times in the late 1840s, amongst numerous other example.

Chapter 2.7: Qualified Determinism

Leveraging off the previous chapter a mathematical notion of qualified determinism is explored here. This

A new function, Dynamic Interaction (DI) that has a 'vertical' and a 'horizontal' component is introduced here. The vertical component is designated as DI_V and the horizontal component as DI_H. Several equations to capture the inherent dynamism that exists in each orientation or state have already been derived in Chapter 2.6. These included equations for the dynamism in the physical, the vital, the mental, and the integral. The derived equations propose a model to give insight into how innovation occurs by changing the fundamental states that an organization is subject to. Several scientists, such as (Prigogine, 1977) and others, are proposing that a system can bifurcate in unpredictable ways to create an emergent property that cannot be predicted. DI is going to propose that in fact there is a 'qualified determinism' as opposed to randomness that occurs.

This qualified determinism is the result of the relative strengths of the levels within the core-matrix identified in the derived equations and summarized by Equation 2.6.7. Hence, the generalized core-matrix, already is the following:

$$\begin{bmatrix} M_3 \rightarrow System_X \\ (\uparrow F \rightarrow I) \\ M_2 \rightarrow S_{System_X} \\ (\uparrow Sig \rightarrow F) \\ M_1 \rightarrow Sig_x \\ (\uparrow > P_{x)} \\ U \rightarrow x_U \end{bmatrix}$$

The application of the vertical component of the new function being proposed, DI_V, to this core matrix will yield the nature or 'strength' of the state (x) or orientation under consideration. If the untransformed or U layer is strongest, implying that the habitual patterns that keep an organization locked into its untransformed way of operation are still very active, then the nature of the output of DI_V, notated by x-state, will be x_U. If the habitual patterns have been overcome then the strength of the x-state increases since it is the dynamics of M_1 or Sig_x that are now active. In this case the x-state will be Sig_x. If the unique 'signature' has become a 'force', then the conditions for activation of M_2 have been put in place and the x-state will be even higher, S_{System_X}. The architectural forces active in M_2 are by definition more powerful than Sig_x that is a derivation of a set of such architectural forces. If the 'force' so acting becomes impersonal so that an organizational ego-state is overcome, then the x-state will have the most strength and is characterized by $System_X$ active at M_3. Hence, DI_V applied to a core-matrix will yield the 'strength' in terms of the x-dynamic that is active. This is illustrated by the following equation, Equation 2.7.1 which can be considered to be a deductive proof in the context of this model:

$$DI_V \begin{bmatrix} M_3 \rightarrow System_X \\ (\uparrow F \rightarrow I) \\ M_2 \rightarrow S_{System_X} \\ (\uparrow Sig \rightarrow F) \\ M_1 \rightarrow Sig_x \\ (\uparrow > P_{x)} \\ U \rightarrow x_U \end{bmatrix} =>$$

$$\text{x-state} \in \left(x_U, Sig_x, S_{System_X}, System_X\right)$$

$$Where: Strength(System_X) > Strength(S_{System_X}) > Strength(Sig_x) > Strength(x_U)$$

Eq 2.7.1: Illustrating Action of Dynamic Interaction – vertical component

What is to be noted here is that while the action of DI_V yields a relative strength and therefore a 'single' value for the core- or x-matrix under consideration yet each x-matrix in itself could have an infinite number of possibilities. This should be clear in looking at how x_U, Sig_x, S_{System_X}, and $System_X$, were initially defined.

Hence, taking the example where x = physical:

$$Physical_U \ni [inertia, lethargy, status\ quo, \dots]$$

As can be seen $Physical_U$, defined in Chapter 2.6, is already an infinite set with qualities similar to the ones already specified.

Similarly, Sig_P, defined in Chapter 2.4, also has an infinite variation:

$$Sig_P = Xa + \overline{Yb_{0-n}} \quad where \begin{bmatrix} X \in [S_{System_{Pr}}] \\ Y \in [S_{System_{Pr}}, S_{System_P}, S_{System_K}, S_{System_N}] \\ a, b\ are\ integers; a > b \end{bmatrix}$$

$S_{System_{Pr}}$, defined in Chapter 2.3, is also an infinite set with forces of the nature specified in the following equation:

$$S_{System_{Pr}} \ni [Service, Perfection, Diligence, Perseverance, \dots]$$

And recall that in Chapter 2.2, $System_{Pr}$ has been defined as:

$$System_{Pr} \equiv TS_{0\rightarrow N}\left[\begin{matrix} Presence \\ \downarrow \\ Opportunity \end{matrix} \begin{bmatrix} P_L & \rightarrow & V_L \\ V_L & \rightarrow & M_L \end{bmatrix} \ \&\ Container_{\begin{bmatrix} System_P \\ System_K \\ System_N \end{bmatrix}}\right]$$

So in essence DI_V is really giving us a summary assessment of the 'level' of the x-matrix under consideration with all its infinite potentiality. An example will follow shortly.

The other component of DI, as suggested earlier in this chapter, is the horizontal component, DI_H. Just as DI_V yields a summary assessment of the level that an x-matrix is operating at, similarly DI_H yields a summary assessment of the direction that a system or organization under consideration is going to continue its development in considering the physical, the vital, the mental, and the integral orientations to be the choices.

Assuming that any organization or system is inherently unique, as this mathematical model proposes, and assuming that the infinite sets of x_U and S_{System_X} applied across the physical, vital, mental, and integral orientations respectively will account for any state that an organization can experience, then at a certain point in time any organization under consideration is going to have a direction-bias in one of the possible physical, vital, mental, or integral directions. Hence, DI_H will yield the summary direction that is going to lead an organization into its future given the current states active in it.

This summary direction is going to be yielded by considering the relative strengths of the separate core x-matrices – the physical, the vital, the mental, the integral - under consideration. The assumption is that there will be one core-matrix that will be stronger than the others.

Hence, as an example, first applying DI_V across all four x-matrices may, for example, yield the following results, with the strongest level within each x-matrix highlighted and bolded:

$$\begin{bmatrix} System_{Pr} \\ \boldsymbol{S_{System_{Pr}}} \\ Sig_P \\ Physical_U \end{bmatrix} \begin{bmatrix} System_P \\ S_{System_P} \\ Sig_V \\ \boldsymbol{Vital_U} \end{bmatrix} \begin{bmatrix} System_K \\ S_{System_K} \\ Sig_M \\ \boldsymbol{Mental_U} \end{bmatrix} \begin{bmatrix} System_N \\ S_{System_N} \\ \boldsymbol{Sig_I} \\ Integral_U \end{bmatrix}$$

Since by definition the strength of $System_x$ is greater than S_{System_x}, which is greater than Sig_x, which is greater than x_U, applying DI_H, as in Equation 2.7.2, across these x-matrices, as in the example following it will then yield the strongest direction, which in this example is the Physical:

$$DI_H\left(\begin{bmatrix} System_{Pr} \\ S_{System_{Pr}} \\ Sig_P \\ Physical_U \end{bmatrix} \begin{bmatrix} System_P \\ S_{System_P} \\ Sig_V \\ Vital_U \end{bmatrix} \begin{bmatrix} System_K \\ S_{System_K} \\ Sig_M \\ Mental_U \end{bmatrix} \begin{bmatrix} System_N \\ S_{System_N} \\ Sig_I \\ Integral_U \end{bmatrix}\right) = Orientation_{Strongest}$$

Eq 2.7.2: Illustrating Action of Dynamic Interaction - horizontal component

Example:

$$DI_H\left(\begin{bmatrix} System_{Pr} \\ \boldsymbol{S_{System_{Pr}}} \\ Sig_P \\ Physical_U \end{bmatrix} \begin{bmatrix} System_P \\ S_{System_P} \\ Sig_V \\ \boldsymbol{Vital_U} \end{bmatrix} \begin{bmatrix} System_K \\ S_{System_K} \\ Sig_M \\ \boldsymbol{Mental_U} \end{bmatrix} \begin{bmatrix} System_N \\ S_{System_N} \\ \boldsymbol{Sig_I} \\ Integral_U \end{bmatrix}\right) = Physical$$

Hence, DI function will yield the following organizational direction, as in Equation 2.7.3, where 'x_matrix' is used interchangeably with 'orientation':

$$Org\_Dir = DI\left(\begin{matrix}\begin{bmatrix}M_3 \rightarrow System_{Pr}\\(\uparrow F \rightarrow I)\\M_2 \rightarrow S_{System_{Pr}}\\(\uparrow Sig \rightarrow F)\\M_1 \rightarrow Sig_P\\(\uparrow > P_{P)}\\U \rightarrow Physical_{\square}\end{bmatrix} & \begin{bmatrix}M_3 \rightarrow System_P\\(\uparrow F \rightarrow I)\\M_2 \rightarrow S_{System_P}\\(\uparrow Sig \rightarrow F)\\M_1 \rightarrow Sig_V\\(\uparrow > P_{V)}\\U \rightarrow Vital_U\end{bmatrix}\\ \begin{bmatrix}M_3 \rightarrow System_S\\(\uparrow F \rightarrow I)\\M_2 \rightarrow S_{System_S}\\(\uparrow Sig \rightarrow F)\\M_1 \rightarrow Sig_M\\(\uparrow > P_{M)}\\U \rightarrow Mental_U\end{bmatrix} & \begin{bmatrix}M_3 \rightarrow System_N\\(\uparrow F \rightarrow I)\\M_2 \rightarrow S_{System_N}\\(\uparrow Sig \rightarrow F)\\M_1 \rightarrow Sig_I\\(\uparrow > P_{I)}\\U \rightarrow Integral_U\end{bmatrix}\end{matrix}\right) \rightarrow$$

$$x\_matrix_{strongest} @ level_{strongest}$$

Eq 2.7.3: Organizational Direction

Generalizing, as in Equation 2.7.4, where Org_Dir is organizational direction:

$$Org\_Dir = DI\left(\begin{bmatrix}M_3 \rightarrow System_X\\(\uparrow F \rightarrow I)\\M_2 \rightarrow S_{System_X}\\(\uparrow Sig \rightarrow F)\\M_1 \rightarrow Sig_x\\(\uparrow > P_{P)}\\U \rightarrow x_U\end{bmatrix}_{x=p,v,m,i}\right) \rightarrow$$

$$x\_matrix_{strongest} @ level_{strongest}$$

Eq 2.7.4: Generalized Equation for Organizational Direction

Hence, this mathematical model is suggesting that any situation, rather than having a random outcome, has a 'qualified deterministic' outcome. In the introduction to his book "Where is Science Going?" (Planck, 1933),

James Murphy points out that the reason Planck spent so much of his time giving lectures on causation was because of the trend of physicists at the time, which has continued to the modern day, to overthrowing the principle of causation following the development of quantum theory, which he felt was misplaced. "Planck would claim", he wrote, "and so would Einstein, that it is not the principle of causation itself which has broken down in modern physics, but rather the traditional formulation of it." Murphy also quotes James Jeans (Jeans, 1932) to suggest the issue associated with causation and determinism: "Einstein showed in 1917 that the theory founded by Planck appeared, at first sight at least, to entail consequences far more revolutionary than mere discontinuity", and here he is referring to the finding that radiant energy is not emitted in a continuous flow, but in integral quantities, or quanta, which can be expressed in integral numbers. Continuing: "It appeared to dethrone the law of causation from the position it had therefore held as guiding the course of the natural world. The old science had confidently proclaimed that nature could follow only one road, the road which was mapped out from the beginning of time to its end by the continuous chain of cause and effect; state A was inevitably succeeded by state B. So far the new science has only been able to say that state A may be followed by state B or C or D or by innumerable other states. It can, it is true, say that B is more likely than C, C than D, and so on; it can even specify the relative probabilities of B, C, and D. But, just because it has to speak in terms of probabilities, it cannot speak with certainty which state will follow which; this is a matter which lies on the knees of the gods – whatever gods there may be."

While under the apparent dynamics at the quantum level there may appear to be randomness and a dethroning of the principle of causation, the notion of a multiplicity of levels, each having its impact on the strength of an orientation and further on the consequent direction from a multiplicity of possible orientations, is being suggested here as determining the direction of any CAS, while still allowing infinite variation in the details that may define its. Hence, the positions of Planck and Einstein are vindicated when considering Equation 11.4.

Further, assuming any CAS where multiple elements are active, connected, interdependent, and emergent, it may be possible to understand, through application of calculus, as to which level is the source for change.

Hence, where N may be source of change, the rate of change of N will resolve into one of P_U, V_U, M_U, I_U, P_T, V_T, M_T, or I_T. This may be summarized by Equation 2.7.5, where y is either U or T:

$$\frac{dN}{dt} \rightarrow \begin{bmatrix} P_U & P_T \\ V_U & V_T \\ M_U & M_T \\ I_U & I_T \end{bmatrix} \rightarrow x_y, where\ y \in (U,T)$$

Eq 2.7.5: Establishing the Nature of the Change

If T, implying that the action of one of the meta-levels has caused transformation, then application of one of the following integrals will determine which level is the likely source for change.

Hence, for M_1 if the integral of $\frac{\partial(x_U \rightarrow x_T)}{\partial t}$ across a limited area 'a' in the vicinity of the change, is greater than some threshold value $Threshold_{Signature}$, then the signature dynamics are likely the source of change. This is summarized by Equation 2.7.6:

$$\int_0^a \frac{\partial(x_U \rightarrow x_T)}{\partial t} > Threshold_{Signature}$$

Eq 2.7.6: Signature Dynamics as the Source of Change

For M_2 if the integral of $\frac{\partial(x_U \rightarrow x_T)}{\partial t}$ across a larger area 'b' extending beyond the vicinity of the change, is greater than some threshold value $Threshold_{Architectural\ Forces}$, then the architectural forces are likely the source of change. This is summarized by Equation 2.7.7:

$$\int_0^b \frac{\partial(x_U \rightarrow x_T)}{\partial t} > Threshold_{ArchitecturalForces}$$

Eq 2.7.7: Architectural Forces as the Source of Change

For M_3 if the double integral of $\frac{\partial(x_U \rightarrow x_T)}{\partial t}$ across the CAS specified by 'A', and across some time 't', is greater than some threshold value $Threshold_{System\ Property}$, then the system properties are likely the source of change. This is summarized by Equation 2.7.8:

$$\int_0^t \int_0^A \frac{\partial(x_U \rightarrow x_T)}{\partial t} > Threshold_{SystemProperty}$$

Eq 2.7.8: System Properties as the Source of Change

Chapter 2.8: Equation for Light-Space-Time Emergence

Equation 2.6.6, the generalized equation for innovation, can be restated as an evolving form true for all time, as in Equation 2.8.1:

$$Innovation_{orientation-x} = \left(\begin{bmatrix} M_3 \rightarrow System_X \\ (\uparrow F \rightarrow I) \\ M_2 \rightarrow S_{System_X} \\ (\uparrow Sig \rightarrow F) \\ M_1 \rightarrow Sig_x \\ (\uparrow > P_x) \\ U \rightarrow x_U \end{bmatrix} TC \rightarrow x_T \text{, } where \begin{bmatrix} x_U \ni [\ldots] \\ x_T \ni [\ldots] \end{bmatrix} \right)_{\langle x_U | x_T \rangle}$$

Eq 2.8.1: Evolving Form of Generalized Equation of Innovation

But further, given that quanta is proposed to be a doorway to deeper worlds of Light, that in fact allow aspects of those worlds or layers to become active at the surface layer U, the question is when have those aspects become active in manifest time. The following suggests when:

- At time, t = 0 seconds, only M_3 is active. Recall that M_3 represents the four-fold reality present in every point-instant.
- At time, $0 \geq t > \infty$, M_2 the set of architectural forces continually gets added to, thereby increasing the size of the sets of forces.
- At time, $0 > t \geq 10^{-36}$ seconds, the equation of Innovation, $Innovation_{orientation-x}$, is such that M_1 also becomes active, while U and TC are suppressed while present. The activation of M_1

begins to result in unique expressions or signatures of the set of architectural forces, and in this case in the reality of the essentially ubiquitous EM Spectrum as a vehicle of the four-fold order that expressed itself in all that existed and in all that unfolded from that point in time on.

- At time, $t \sim 10^{-10}$ seconds, fundamental particles emerge as an essential material basis of the four architectural forces that frame all further development. As in the case of the EM Spectrum this implies the activity of M_1.
- At time, $t \sim 3 \times 10^5$ years light atoms emerge, and at time $t \sim 10^9$ years heavier atoms in the stars emerge. These also imply the continued activity of M_1.
- At time, $t \sim 13.8 \times 10^9$ years, the fourth clear expression of the same four-fold order as the bases of an even more complex organization, that of cellular life and all that is founded on it becomes clear. This time-point will be represented by the notation $t \sim E_{Cell}$, where 'E' stands for emergence. This too implies the activity of M_1. But further, TC becomes active here, with an implicit direction of operation from M_3 to U. Note that the sets of architectural forces specified by M_2 continue to increase the number of elements they comprise of as the complex interaction between the layers continues.
- At time $t > 13.8 \times 10^9$ years, human-beings, and more complex social organizations emerge. Here TC acts with an implicit direction of operation from U to M_3. This time-point will be represented by $t \sim E_{Human}$.

Based on the aforementioned description Equation 2.8.2 for Emergence true of any space-time scale may be generalized as the following:

$$
Emergence_{space-time} = \left| \begin{array}{c}
\begin{bmatrix} M_3 \rightarrow System_X \\ (\uparrow F \rightarrow I) \\ M_2 \rightarrow S_{System_X} \\ (\uparrow Sig \rightarrow F) \\ M_1 \rightarrow Sig_x \\ (\uparrow > P_{x)} \\ U \rightarrow x_U \end{bmatrix}_{Space} \\
\left[U \rightarrow \begin{array}{c} M_3 : All\ t \\ \downarrow \\ M_2 : 0 \geq \mathrm{t} > \infty \\ \downarrow \\ M_1 : 0 > \mathrm{t} > \infty \\ \downarrow \\ \mathrm{t} \sim E_{Cell}; \mathrm{TC}: M_3 \rightarrow \mathrm{U} \\ \mathrm{t} \sim E_{Human}; \mathrm{TC}: U \rightarrow M_3 \end{array} \right]_{Time} \\
TC \rightarrow x_T, where \begin{bmatrix} x_U \ni [\ldots] \\ x_T \ni [\ldots] \end{bmatrix}
\end{array} \right|_{\langle x_U | x_T \rangle}
$$

Eq 2.8.2: Space-Time Emergence

An implication of this equation, brought out more explicitly through the elaboration of the 'Time' component, is that the layers U, M_1, M_2, and M_3 exist simultaneously. Adding the Light-Matrix derived in Chapter 2.1 enhances Equation 2.8.2 to the Light-Space-Time Emergence form as represented by Equation 2.8.3:

$$
Emergence_{light-space-time} = \left| \begin{array}{cc} \begin{bmatrix} C_{\infty}: [Pr, Po, K, H] \\ (\downarrow R_{C_K} = R_{C_{\infty}}) \\ C_K: [S_{Pr}, S_{Po}, S_K, S_H] \\ (\downarrow R_{C_N} = R_{C_K}) \\ C_{N}: f(S_{Pr}\ x\ S_{Po}\ x\ S_K\ x\ S_H) \\ (\downarrow R_{C_U} = R_{C_N}) \\ C_U: [P, V, M, C] \end{bmatrix}_{Light} & \begin{bmatrix} M_3 \to System_X \\ (\uparrow F \to I) \\ M_2 \to S_{System_X} \\ (\uparrow Sig \to F) \\ M_1 \to Sig_x \\ (\uparrow > P_x) \\ U \to x_U \end{bmatrix}_{Space} \\ \begin{bmatrix} M_3 : All\ t \\ \downarrow \\ M_2 : 0 \geq t > \infty \\ \downarrow \\ M_1 : 0 > t > \infty \\ \downarrow \\ U \to \begin{array}{c} t \sim E_{Cell};\ TC: M_3 \to U \\ t \sim E_{Human};\ TC: U \to M_3 \end{array} \end{bmatrix}_{Time} & TC \to x_T \end{array} \right|_{\langle x_U | x_T \rangle}
$$

Eq 2.8.3: Light-Space-Time Emergence ʻ

(2.8.3) will be further explored in Section 6.

Chapter 2.9: Speed of Light and Quanta

As discussed conceptually in Section 1 and also mathematically in Chapter 2.1, since c is finite and therefore there is past, present, and future implied by it, this implies that at U a point has to become quanta. This is implicit in the notion of finiteness. Since light takes a finite amount of time to get from A to B, a "unit" of light will require a finite time to traverse that. Quanta at the subatomic level can be thought of as related to this finite time and distance for a unit of light to be expressed.

Planck's discovery that energy at the subatomic level acts as quanta therefore makes sense. It is to be noted though that Planck's treatment of quanta was more as a mathematical convenience that allowed the derivation of an equation that explained the curve of radiation wavelengths at varying temperatures of a heated black-body (Isaacson, 2008). Einstein though postulated quanta as a fundamental property of light itself, rather than as something that arose in the interaction of light with matter as Planck thought. Einstein's theory produced a law of the photoelectric effect where the energy of emitted electrons would depend on the frequency of light. Einstein received the Nobel Prize for this discovery (Isaacson, 2008).

Summarizing, if c is the upper limit of the layer U, then it makes sense that the lower limit h (Planck's constant) should be inversely proportional to c. Hence:

$$h \propto \frac{1}{c}$$

This relationship is substantiated by combining two well-known equations: the first is the electromagnetic equation connecting speed of light with wavelength and frequency, and the second is Einstein's photoelectric equation connecting energy with frequency of light:

(1) $C = \nu\lambda$
(2) $E = h\nu$

Yields:

$$h = \frac{E\lambda}{C}$$

About h, H.A. Lorentz the Dutch scientist has commented in The Science of Nature (Lorentz, 1925): "We have now advanced so far that this constant not only furnishes the basis for explaining the intensity of radiation and the wavelength for which it represents a maximum, but also for interpreting the quantitative relations existing in several other cases among the many physical quantities it determines. I shall mention a few only, namely the specific heat of solids, the photo-chemical effects of light, the orbits of electrons in the atom, the wavelengths of the lines of the spectrum, the frequency of the Roentgen rays which are produced by the impact of electrons of given velocity, the velocity with which gas molecules can rotate, and also the distances between the particles which make up a crystal. It is no exaggeration to say that in our picture of nature nowadays it is the quantum conditions that hold matter together and prevent it from completely losing its energy by radiation."

So just as c sets up the past-present-future experience and reality of U, h suggests that this experience will take place in shells of matter. In the absence of the limit h, as pointed out by Lorentz, only radiation, and no matter would exist. This reinforces the notion of the four-foldness implicit in the nature of light as already discussed in Section 1.

The suggested variance in the speed of light by meta-layer may also throw some further light on the quantum realm. First, summarizing:

1. At U the speed of light in a vacuum, c_U, is finite at 186,000 miles/sec. This finiteness creates the reality and experience of past-present-future, and further a sense of fragmentation and separation. Further, assuming that c_U is a fundamental upper-limit at U, the inverse of it , $\frac{1}{c_U}$, must define some fundamental lower limit at U. This is indeed the case as Planck's constant, h, is proportional to this. 'h' allows for matter to be sustained, as it fundamentally limits the dispersion of energy as suggested by Lorentz.
2. At M_3, the speed of light, c_{M_3}, is suggested as being ∞ miles/sec. This allows a reality of 'oneness' and the possibility of a suggested four-fold intelligence existing in every point-instant of space-time as suggested in Chapter 1.2.
3. As suggested in Chapter 3.1 the quantum world, here designated by Q, because it is at boundary of U, accesses and interrelates with the meta-levels. As such, the speed of light, c_Q, will appear as a hybrid as in the following figure. Note though that it is really the speed of light at the native or resident layer that becomes active, and that this is simply being represented as c_Q for convenience:

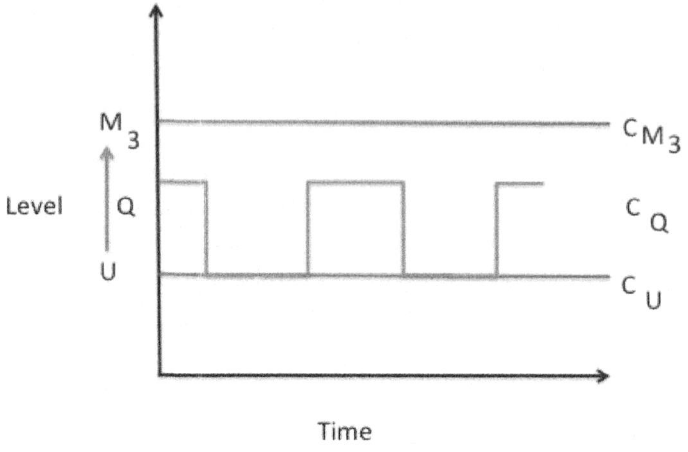

Figure 2.9.1 Speed of Light at Quantum Level

Note that research on the speed of light also indicates that it may go faster than c_U. While the speeds suggested currently through experimental research, and summarized below, may be only incrementally higher than c_U the notion that c_U can be exceeded appears to be put in place:

1. The Heisenberg Uncertainty Principle (to be discussed subsequently in Chapter 2.11) already suggests that photons can travel at any speed, even exceeding c_U, for short periods.
2. Notion of different space-time realities, also known as meta-levels, suggests that light can travel differently in a layer different from the four-dimensional space-time that apparently defines our observable world (Hawking, 1988)
3. In his book QED Feynman (Feynman, 1985) says "...there is also an amplitude for light to go faster (or slower) than the conventional speed of light. You found out in the last lecture that light doesn't go only in straight lines; now, you find out that it doesn't go only at the speed of light! It

may surprise you that there is an amplitude for a photon to go at speeds faster or slower than the conventional speed, c." In research conducted at Humboldt University (Chown, 1990), Scharnhorst has made calculations using the theory of quantum electrodynamics to reveal the possible existence of "faster-than-light" photons. This is known as the Scharnhorst effect.

4. As reviewed in Chapter 2.1, Perkowitz makes the point that the theory of relativity does not disallow particles already moving at c or greater.

The point is that the reality at Q is going to be different than the reality at U. This should be apparent from considering the relation of c_X to the consequent reality as discussed in Section 1. In Q the fundamental lower limit, h, which allows matter to sustain itself, is itself going to fluctuate. Hence, as X tends to M_3, c_X will tend to infinity, and h will become a fraction of itself. As it becomes a fraction of itself the quantization effect will be lowered, and matter will get dispersed more and more easily to in effect take on a wave-like appearance.

Chapter 2.10: Interpreting Schrodinger's Equation

Schrodinger's equation, which seeks to model how a quantum state of a quantum system changes with time, or in other words seeks to model matter as a wave rather than as a particle (Stewart, 2012), is depicted in Equation 2.10.1:

$$i\frac{h}{2\pi}\frac{\partial}{\partial x}\psi = \hat{H}\psi$$

Eq 2.10.1: Schrodinger's Equation

ψ depicts a wave form and can be thought of as a probable cloud of possible states. $\hat{H}$ is the Hamiltonian operator which is a focusing function, and in its essence what the equation may be suggesting is that the way a wave form changes over time is equivalent to some expressible state of the possibilities inherent in the cloud of possible states.

But the cloud of possible states is another way of saying that behind the layer U form is represented in another way than at U. If the existence of the meta-levels, and in this case, of Sig_x at M_1, is considered possible, then it is far more reasonable to admit that form is configured by function and the very dynamics of what may appear to have been random, as in many interpretations of quantum phenomena, may now appear to be far more logical. There is now more *context* to interpreting observation at the quantum level.

Interestingly Schrodinger himself had misgivings about the applicability of this equation that seemed to apply at the quantum level, to the macro-world (Stewart, 2012). To bring his misgivings to light he invented a thought experiment concerning a cat. This cat would be in a superposed state in a quantum black-box. A radioactive particle, a decaying-particle detector, and a flask of poison, were the other inhabitants of the black-box. At some point the particle will decay, be detected, and as in the thought-experiment at that point, triggered by the decaying particle, the poison in the flask would be released. The cat would then die. But in the meanwhile the cat would be in a superposed states of being both dead and alive. Only when the box was opened would the wave function collapse and a single definite state emerge.

Schrodinger was hoping to highlight the absurdity of the application of having a cat in both a dead and alive state at the macro-level. Instead physicists found this thought experiment to be sensible, even at the macro-level, and began to generalize the findings based on this. Hence, for example the idea of superposition at the physical level began to be thought of as real. It is interesting to note that in his lectures on Schrodinger's equation Feynman (Gottlieb, 2013) has stated "Where did he get that [equation] from? Nowhere. It is not possible to derive it from anything you know. It came out of the mind of Schrödinger".

Considering Schrodinger's equation in the light of the discussion on Q in Chapter 2.9. 'i' is a complex number and suggests the interplay of two dimensions, one being real, and one being 'imaginary'. But the 'imaginary' dimension could be thought of as none other than the meta-levels implicit in this mathematical model, and suggested to be real at Q. Further, $\frac{h}{2\pi}$ is in line with the suggestion just made that h will have to become a fraction of itself as c increases. Hence, the change in the wave function, $\frac{\partial}{\partial x}\psi$, is intimately related to i and $\frac{h}{2\pi}$, and perhaps only makes sense when considered in the context of i x $\frac{h}{2\pi}$ x $\frac{\partial}{\partial x}\psi$, which has to be the case when dealing with the integration of dynamics of multiple levels.

Further, the change in the wave function, $\frac{\partial}{\partial x}\psi$, is related to $\hat{H}\psi$, and suggests that there is some system "energy", represented by the Hamiltonian, $\hat{H}$, that when applied to the existing wave, ψ, will indicate how the wave will be expressed going forward.

But as discussed, at Q the dynamics of Sig_X become real, and in fact is a fundamental organizing principle for all organization at U, and starting at the dimension of h.

It may be the case that as the level of complexity of organization at the micro-level increases, as in from quantum, to atomic, to cellular, $\hat{H}$, itself complexifies as it houses further nuances or "superpositions" of Sig_X. Sig_X may be thought of as having more components as the complexity at such micro-levels of organization increases. Further, since it is being proposed here that all complexity or innovation is due to the integration of the dynamics of meta-levels with U, the representation of wave functions or ψ at the boundary world, Q, may be enhanced by consideration of Sig_X. Hence, $\hat{H}$ may be further qualified by notating $\hat{H}_Q$ or $\hat{H}_A$ or $\hat{H}_C$ for quantum, atomic, and cellular respectively, as per the following inequality in Equation 2.10.2:

$$\hat{H}_Q < \hat{H}_A < \hat{H}_C \text{ (where: } \hat{H}_X \propto Energy_{Sig_X})$$

Eq 2.10.2: Hamiltonian Inequality

This inequality is specified by the number of components of Sig_X which it is assumed will have a direct effect on potential and kinetic energy of the wave-system (which are considered to be how $\hat{H}$ is measured) as in Equation 2.10.3:

$$Energy_{sig_Q} < Energy_{sig_A} < Energy_{sig_C}$$

Eq 2.10.3: Quantum, Atomic, Cellular Energy Inequality

Similarly the following wave equations for quantum, atomic, and cellular respectively, as in Equations 2.10.4, 2.10.5, and 2.10.6 may be distinguished:

$$i\frac{h}{2\pi}\frac{\partial}{\partial x}\psi_Q = \hat{H}_Q\psi_Q$$

Eq 2.10.4: Hamiltonian at Quantum Level

$$i\frac{h}{2\pi}\frac{\partial}{\partial x}\psi_A = \hat{H}_A\psi_A$$

Eq 2.10.5: Hamiltonian at Atomic Level

$$i\frac{h}{2\pi}\frac{\partial}{\partial x}\psi_C = \hat{H}_C\psi_C$$

Eq 2.10.6: Hamiltonian at Cellular Level

Chapter 2.11: Interpreting Heisenberg's Uncertainty Principle

In his book, The Little Book of String Theory, Princeton University's Gubser (Gubser, 2010) describes the effect on approaching absolute zero temperature on molecules. He takes the example of water molecules and relates that one cannot make the water molecules colder than absolute zero, -273.15 Celsius, because there is no more thermal energy to suck out at that temperature. However, quantum uncertainty, the phenomenon which relates the momentum and location of electrons in atoms necessitates that the water molecules will still vibrate. Gubser suggests this by considering Heisenberg's uncertainty relation, reproduced in Equation 2.11.1:

$$\Delta p \; X \; \Delta x \; \geq \frac{h}{4\pi}$$

Eq 2.11.1: Heisenberg's Uncertainty Relation

In Equation 2.11.1, Δp is the uncertainty in a particle's momentum, Δx is the uncertainty in the particle's location, and h is the Planck's constant. In frozen water crystals it is precisely known where the water molecules are, and therefore Δx is fairly small. This means that Δp has to be considerably larger, and therefore that the water molecules are still vibrating even though they are at absolute zero. This innate vibration, known as 'quantum zero-point' energy, expresses the phenomenon of quantum fluctuations.

The Planck's constant order of magnitude (10^{-34}) though, suggests the boundary between U and $M_1$and the quantum fluctuations, the uncertainty relation, and the quantum zero-point energy could be an expression of the essential Signature function, Sig_x, that is posited as a key formative force behind organization at U. In this interpretation the thermal energy describes the essential energy at U, while the uncertainty relation may suggest the phenomenon of innovation or function-precipitation "physically" linking M_1 and U. In this case it may be suggested that integration of meta-levels with the surface level, I_U^M, is indicated by the uncertainty relation, as in Equation 2.11.2:

$$I_U^M \rightarrow \Delta p \; X \; \Delta x \geq \frac{h}{4\pi}$$

Eq 2.11.2: Integration of Levels (Leveraging Heisenberg's Uncertainty Relation)

But further, it may also be suggested that the uncertainty principle itself is only valid at U, and that, because of the finiteness of c. This finiteness as already suggested implies h, which implies that if the position of a particle is going to be observed by shining light on it, the light has to have at least a quantum of energy. But to determine the position of a particle accurately light of a shorter wavelength would have to be used (Hawking, 1988) which would have to have a minimum amount of energy, which in turn would interfere with the velocity and hence momentum of the particle. The uncertainty in measuring the momentum could therefore be thought of as a consequence of the finiteness of the speed of light, c.

If c_U were to approach c_Q though, the quantum would be smaller and the uncertainty in measuring position or momentum would be reduced. At c_{M_3} there would be no uncertainty since light would accurately tell both position and momentum definitively.

Hence, the uncertainty principle may be further qualified, as in Equations 2.11.3, 2.11.4, and 2.11.5:

$$@C_U: \Delta p \; X \; \Delta x \geq \frac{h}{4\pi}$$

Eq 2.11.3: Uncertainty Principle at U

$$@C_Q: \Delta p \; X \; \Delta x \rightarrow 0$$

Eq 2.11.4: Uncertainty Principle at Q

$$@C_{M_3}: \Delta p \; X \; \Delta x = 0$$

Eq 2.11.5: Uncertainty Principle at M3

The notion of position and momentum becoming finite at U also may imply that space, time, and quanta are emergent rather than absolute properties, as also suggested in Section 1 and specifically in Chapter 1.6. This is also the conclusion of Arkani-Hamed of the Institute of Advanced Studies in the following thought experiment (Wolchover, 2013):

'Locality says that particles interact at points in space-time. But suppose you want to inspect space-time very closely. Probing smaller and smaller distance scales requires ever higher energies, but at a certain scale, called the Planck length, the picture gets blurry: So much energy must be concentrated into such a small region that the energy collapses the region into a black hole, making it impossible to inspect. "There's no way of measuring space and time separations once they are smaller than the Planck length," said Arkani-Hamed. "So we imagine space-time is a continuous thing, but because it's impossible to talk sharply about that thing, then that suggests it must not be fundamental — it must be emergent."

Unitarity says the quantum mechanical probabilities of all possible outcomes of a particle interaction must sum to one. To prove it, one would have to observe the same interaction over and over and count the frequencies of the different outcomes. Doing this to perfect accuracy would require an infinite number of observations using an infinitely large measuring apparatus, but the latter would again cause gravitational collapse into a black hole. In finite regions of the universe unitarity can therefore only be approximately known.'

Chapter 2.12: A Deeper Look at Quantization of Space, Time, Matter, and Gravity

The Light-Space-Time Emergence equation (2.8.3) derived in Chapter 2.8, and reproduced here for convenience, offers some insight into the process of quantization. Reproducing the equation:

$$Emergence_{light-space-time} = \left\| \begin{matrix} \begin{bmatrix} C_\infty: [Pr, Po, K, H] \\ (\downarrow R_{C_K} = R_{C_\infty}) \\ C_K: [S_{Pr}, S_{Po}, S_K, S_H] \\ (\downarrow R_{C_N} = R_{C_K}) \\ C_{N:}\ f(S_{Pr}\ x\ S_{Po}\ x\ S_K\ x\ S_H) \\ (\downarrow R_{C_U} = R_{C_{N)}} \\ C_U: [P, V, M, C] \end{bmatrix}_{Light} & \begin{bmatrix} M_3 \rightarrow System_X \\ (\uparrow F \rightarrow I) \\ M_2 \rightarrow S_{System_X} \\ (\uparrow Sig \rightarrow F) \\ M_1 \rightarrow Sig_x \\ (\uparrow > P_{x)} \\ U \rightarrow x_U \end{bmatrix}_{Space} \\ \begin{bmatrix} M_3 : All\ t \\ \downarrow \\ M_2 : 0 \geq t > \infty \\ \downarrow \\ M_1 : 0 > t > \infty \\ \downarrow \\ U \rightarrow \begin{matrix} t \sim E_{Cell};\ TC: M_3 \rightarrow U \\ t \sim E_{Human};\ TC: U \rightarrow M_3 \end{matrix} \end{bmatrix}_{Time} & TC \rightarrow x_T \end{matrix} \right|_{\langle x_U | x_T \rangle}$$

The Light-Matrix suggests that there are particular kinds of quantization that could occur. Along the vertical realm these can be thought of as inter-relating one layer of the matrix with the previous layers. Hence the kind of quantization that is relevant to the physical layer, U, would relate C_U with C_X. This can be represented by h_U, where h stands for Planck's constant. Note that in this model there are other fundamental quantization possible that may inter-relate a specific light-layer with others above it.

In looking at the Light-Matrix it is also clear that each subsequent layer below the top layer, describing the reality set up by an infinite speed, is emergent. Hence space, time, matter, and gravity, can be thought of as emergent phenomena. The emergence itself can be thought of as a function of Light-Matrix. At layer U, and as discussed in Chapter 1.6, Space is related to Light's property of Knowledge, Time is related to Light's property of Power, Matter or Energy is related to Light's property of Presence, and Gravity is related to Light's Property of Harmony or Nurturing. Each of these will emerge in a particular way and it is possible that there are multiple h's.

Hence, the Planck's constant for matter or energy, which we are already familiar with from past discoveries in science, can be depicted as h_{UPr}, since it is related to Presence (Pr). But similarly, quantization for space, time, and gravity could potentially be governed by other similar constants, referred to as h_{UK}, h_{UP}, and h_{UH}, and related to Knowledge (K), Power (P), and Harmony (H), respectively. Since the relationship between these constants, in absolute terms, is uncertain, this can be represented by the equality-inequality as depicted by Equation 2.12.1:

$$h_{UPr} \leq\geq h_{UK} \leq\geq h_{UP} \leq\geq h_{UH}$$

Eq 2.12.1: Equality-Inequality Relationship Between Different 'Planck' Constants

Further, the quantization-window, as it were, potentially allows the precipitation of, or inter-relation with, or creation of a cohesive and compelling meta-function or signature as modeled by the C_N/M_1 layer. This quantization-window is positioned as being key in allowing a phenomenon of quantum-certainty to occur and will be explored in greater detail in the subsequent chapters on the working of quantization using history as an example. The 'Time' matrix, in (2.8.3) reproduced above, indicates the general default direction under which such quantum-certainty can occur at Layer U. Generally at the pre-human level it may proceed more automatically by the dynamics of the system itself. Beyond this level it will generally occur through the action of a cohesive will. The opening of this quantization-window is modeled by the 'Space' matrix and will happen when habitual patterns are overcome so that 'will' becomes cohesive.

The specific quantization that are occurring along the space, time, energy/matter, and gravity dimensions are modeled by the following equations, that are all based on the Sig_X equations derived in Chapter 2.4. As a reminder (2.4.5) is reproduced here for convenience:

$$Sig = Xa + \overline{Yb_{0-n}}$$

$$where: \begin{bmatrix} X \in [S_{System_{Pr}}, S_{System_P}, S_{System_K}, S_{System_N}] \\ Y \in [S_{System_{Pr}}, S_{System_P}, S_{System_K}, S_{System_N}] \\ a, b \text{ are integers}; a > b \end{bmatrix}$$

The Structure of Space

The structure of Space, which holds the seeds of knowledge of all that will emerge, hence, is modeled in the following way as specified by Equation 2.12.2:

$$Space_{quantization} = h_{UK}(Xa + \overline{Yb_{0-n}})$$

$$where: \begin{bmatrix} X \in [S_{System_K}] \\ Y \in [S_{System_{Pr}}, S_{System_P}, S_{System_K}, S_{System_N}] \\ a, b \text{ are integers}; a > b \end{bmatrix}$$

Eq 2.12.2: Space Quantization

In this model 'space' as an emergent property of Light, is structured by infinite seeds of knowledge. But because the emergence is taking place in a layer of reality generally itself structured by a finite speed of light, c, it has to be quantized. The quantization assures that the knowledge is not dissipated, but can accumulate to create seeds, and therefore, the structure of space itself. It is assumed that there is some kind of 'Planck's constant' in effect, that is modeled as being specific to the way knowledge or space may be quantized – hence, h_{UK}.

The Structure of Time

The structure of Time, which holds the inevitability of the seeds or knowledge emerging in a phased maturity, hence, is modeled in the following way as specified by Equation 2.12.3:

$$Time_{quantization} = h_{UP}\left(Xa + \overline{Yb_{0-n}}\right)$$

$$where: \begin{bmatrix} X \in [S_{System_P}] \\ Y \in [S_{System_{Pr}}, S_{System_P}, S_{System_K}, S_{System_N}] \\ a, b \text{ are integers}; a > b \end{bmatrix}$$

Eq 2.12.3: Time Quantization

In this model 'time' as an emergent property of Light, is structured by an inevitable process of maturity, which due to its inevitability, is related to power. But because the emergence is taking place in a layer of reality generally itself structured by a finite speed of light, c, it has to be quantized. The quantization assures that the power is not dissipated, but can accumulate to express phased maturity, and therefore, the structure of time itself. It is assumed that there is some kind of 'Planck's constant' in effect, that is modeled as being specific to the way power or time may be quantized – hence, h_{UP}.

Energy & Matter

Energy, which through a process of containment can result in Matter, hence, is modeled in the following way as specified by Equation 2.12.4:

$$Energy_{quantization} = h_{UPr}(Xa + \overline{Yb_{0-n}})$$

$$where: \begin{bmatrix} X \in [S_{System_{Pr}}] \\ Y \in [S_{System_{Pr}}, S_{System_{P}}, S_{System_{K}}, S_{System_{N}}] \\ a, b \text{ are integers}; a > b \end{bmatrix}$$

Eq 2.12.4: Energy Quantization

In this model 'energy' as an emergent property of Light, results in the reality of matter. But because the emergence is taking place in a layer of reality generally itself structured by a finite speed of light, c, it has to be quantized. The quantization assures that the energy is not dissipated, but can accumulate to create matter. Planck's constant is referred to as - h_{UPr}.

Gravity

Gravity, which holds seemingly distinct objects in the layer of reality created by c together in a harmony, hence, is modeled in the following way as specified by Equation 2.12.5:

$$Gravity_{quantization} = h_{UH}(Xa + \overline{Yb_{0-n}})$$

$$where: \begin{bmatrix} X \in [S_{System_{N}}] \\ Y \in [S_{System_{Pr}}, S_{System_{P}}, S_{System_{K}}, S_{System_{N}}] \\ a, b \text{ are integers}; a > b \end{bmatrix}$$

Eq 2.12.5: Gravity Quantization

In this model 'gravity' as an emergent property of Light, results in a harmonious collectivity of seemingly independent objects. But because the emergence is taking place in a layer of reality generally itself structured by a finite speed of light, c, it has to be quantized. The quantization assures that the harmony is not dissipated, but can accumulate to express more and more complex collectivities on large-scale. It is assumed that there is some kind of 'Planck's constant' in effect, that is modeled as being specific to the way harmony or time may be quantized – hence, h_{UH}.

SECTION 3: ELECTROMAGNETIC SPECTRUM

The electromagnetic spectrum is a technical way to refer to light, essentially because amongst its range of properties it displays a tangible and simultaneous electric and magnetic or "electromagnetic" reality as well. So as light becomes more concrete to us or as the possibilities within it begin to emerge, one of the first forms it takes is as the electromagnetic spectrum. It must be the case that the previously surfaced properties of light – Presence, Power, Knowledge, and Harmony – emerge so as to define the very architecture of the electromagnetic spectrum.

Let us look at this in more detail.

But further, Sections 3, 4, and 5 lead up to Section 6, which considers the complexification of four-foldness of increase in functional-richness in the universe.

Chapter 3.1: Harmony

To begin with, we had already looked at how the speed of light sets up the nature of reality because of its speed. So for light traveling at c, past, present, future, and the notion of separation due to creation of islands of matter is the reality. It seems then that c architects the possibility of interaction in our system and can therefore be thought of a projection from the property of Harmony to create the basis for a matter-based harmony. So being, it can be suggested that the nature of the resultant interactions allows matter-based organizations, regardless of scale, to come into their own, to grow into their boundaries, and to form bonds based on the sense of being separated from other perceived organizations. This notion of forming bonds seems also to be related to the "magnetic" in electromagnetic.

In equation form, as in Equation 3.1.1 it would therefore be possible to specify the nature of reality so set up by the electromagnetic spectrum (EM_Spectrum) moving at c:

$$EM\_Spectrum_{Speed} = Xa + \overline{Yb_{0-n}} \text{ where } \begin{bmatrix} X \in [S_{System_N}] \\ Y \in [S_{System_{Pr}}, S_{System_P}, S_{System_K}, S_{System_N}] \\ a, b \text{ are integers}; a > b \end{bmatrix}$$

Eq 3.1.1: Speed of EM Spectrum

The notion of 'connection', 'growing into one's own', 'form bonds', amongst other attributes of such a reality can be seen as elements of the four sets of architectural forces. Hence 3.1.1 suggests the mathematical equation that so defines the nature of reality due to the speed of c of the electromagnetic spectrum.

This also implies that as the speed of c changes, the nature of relationship in such a reality will also change. Perhaps it is that there are several different relationships, or types of harmonies possible.

Leveraging the previously defined inequality for c_X,

$Inequality_C$: $c_U < c_{M_1} < c_{M_2} < c_{M_3}$

It may be suggested that the nature of the 'layer' will change. This is depicted in Equations 3.1.2 through 3.1.5:

$$[Nature(U)]_{c_U} = Xa + \overline{Yb_{0-n}} \quad where \begin{bmatrix} X \in [S_{System_N}] \\ Y \in [S_{System_{Pr}}, S_{System_P}, S_{System_K}, S_{System_N}] \\ a, b \text{ are integers}; a > b \end{bmatrix}$$

Eq 3.1.2: Nature of U as a Function of c(U)

$$[Nature(M_1)]_{c_{M_1}} = Xa + \overline{Yb_{0-n}} \quad where \begin{bmatrix} X \in [S_{System_N}] \\ Y \in [S_{System_{Pr}}, S_{System_P}, S_{System_K}, S_{System_N}] \\ a, b \text{ are integers}; a > b \end{bmatrix}$$

Eq 3.1.3: Nature of M1 as a Function of c(M1)

$$[Nature(M_2)]_{CM_2} = Xa + \overline{Yb_{0-n}} \quad where \begin{bmatrix} X \in [S_{System_N}] \\ Y \in [S_{System_{Pr}}, S_{System_P}, S_{System_K}, S_{System_N}] \\ a, b\ are\ integers; a > b \end{bmatrix}$$

Eq 3.1.4: Nature of M2 as a Function of c(M2)

$$[Nature(M_3)]_{CM_3} = Xa + \overline{Yb_{0-n}} \quad where \begin{bmatrix} X \in [S_{System_N}] \\ Y \in [S_{System_{Pr}}, S_{System_P}, S_{System_K}, S_{System_N}] \\ a, b\ are\ integers; a > b \end{bmatrix}$$

Eq 3.1.5: Nature of M3 as a Function of c(M3)

In every one of the preceding cases what is to be noticed is that the primary element X derives from the architectural set of nurturing, S_{System_N}.

Chapter 3.2: Knowledge

The electromagnetic spectrum contains a range of waves embedded in it. These waves enable many different applications with practical utility apparent in every day life. So for example there is a region in the electromagnetic spectrum that we call radio waves, and others that we know as microwave, infrared visible light, ultraviolet, x-rays and gamma rays that each make possible many technologies that we use every single day. These ranges essentially code a range of technological-possibility, and therefore the wave-range within the electromagnetic spectrum can be thought of as encoding or of expressing some kind of precipitation or projection or emergence of the property of Knowledge. Or put another way, the property of Knowledge that is found in light, emerges as the wave-range implicit in the electromagnetic spectrum.

The EM spectrum itself, with its vast range of natures from gamma rays through visible light through radio waves, with its implicitness of time-space possibility as suggested by frequency (ν) and wavelength (λ), may be thought of as an arrangement of archetypes of what is possible in systems, and therefore is perhaps a precipitation of system-knowledge, as in Equation 3.2.1:

$$EM\_Spectrum_{Structure} = Xa + \overline{Yb_{0-n}} \ where \begin{bmatrix} X \in [S_{System_K}] \\ Y \in [S_{System_{Pr}}, S_{System_P}, S_{System_K}, S_{System_N}] \\ a, b \ are \ integers; a > b \end{bmatrix}$$

Eq 3.2.1: Structure of EM Spectrum

What this also implies is that the significance or intent of the different types of waves that exist can also be expressed by this general equation where the X and Y elements will vary. What precisely these elements are will need to be worked out. A few representative equations, Equations 3.2.2 through 3.2.5 follow:

$$Gamma\ Rays\ _{Intent} = Xa + \overline{Yb_{0-n}}\ where \begin{bmatrix} X \in [S_{System_K}] \\ Y \in [S_{System_{Pr}}, S_{System_P}, S_{System_K}, S_{System_N}] \\ a, b\ are\ integers; a > b \end{bmatrix}$$

Eq 3.2.2: Gamma Rays Intent

Knowing that some of the applications of Gamma Rays are in sterilizing and radiotherapy, these can be attributed as elements to the architectural sets. The totality of the use of can be thought of as the 'intent' of Gamma Rays. An understanding of the uses will allow a full set of the secondary elements to be mapped thus allowing (3.2.2) to be reverse-engineered.

Similarly each of the archetypes present in the electromagnetic spectrum can be mapped out. Sample mapping of intent include x-rays, infrared, and microwaves:

$$X - Rays_{Intent} = Xa + \overline{Yb_{0-n}}\ where \begin{bmatrix} X \in [S_{System_K}] \\ Y \in [S_{System_{Pr}}, S_{System_P}, S_{System_K}, S_{System_N}] \\ a, b\ are\ integers; a > b \end{bmatrix}$$

Eq 3.2.3: X-Rays Intent

$$Infrared_{Intent} = Xa + \overline{Yb_{0-n}} \quad where \begin{bmatrix} X \in [S_{System_K}] \\ Y \in [S_{System_{Pr}}, S_{System_P}, S_{System_K}, S_{System_N}] \\ a, b \; are \; integers; a > b \end{bmatrix}$$

Eq 3.2.4: Infrared Intent

$$Microwaves_{\,Intent} = Xa + \overline{Yb_{0-n}} \quad where \begin{bmatrix} X \in [S_{System_K}] \\ Y \in [S_{System_{Pr}}, S_{System_P}, S_{System_K}, S_{System_N}] \\ a, b \; are \; integers; a > b \end{bmatrix}$$

Eq 3.2.5: Microwaves Intent

Chapter 3.3: Power

The range of different wave-types or wavelengths implicit in the electromagnetic spectrum also moves with different frequencies. Since the speed of light is a constant, and it is known that speed is the product of frequency and wavelength, the greater the wavelength the lower will be the frequency of the wave-type. Conversely the less the wavelength the higher will be the frequency of the wave-type. And energy or power of a wave-type will depend on its frequency.

Hence, gamma rays that have a lower wavelength will have a higher frequency, and higher energy associated with it. Radio waves on the other hand that have a higher wavelength will have a lower frequency and therefore lower energy associated with it. So implicit in the electromagnetic spectrum is a gradient of energy. But it is also known that the penetration power is dependent on frequency. Therefore, the higher the frequency or energy, the higher will be the power of the wave-type. Another way to say this is that the property of Power in Light emerges as the energy-gradient in the electromagnetic spectrum. This Power aspect, by the way, seems to have been captured by the "electro" in electromagnetic.

The energy-gradient implicit in the EM spectrum suggests the power and energy with which knowledge moves and is perhaps a precipitation of system-power, as in Equation 3.3.1:

$$EM\_Spectrum_{Energy} = Xa + \overline{Yb_{0-n}} \quad where \begin{bmatrix} X \in [S_{System_P}] \\ Y \in [S_{System_{Pr}}, S_{System_P}, S_{System_K}, S_{System_N}] \\ a, b \; are \; integers; a > b \end{bmatrix}$$

Eq 3.3.1: EM Spectrum Energy

This energy as suggested is directly proportional to the frequency, of which there is an infinite range as predicted by Maxwell's equations. Different frequencies have different penetration profiles (HyperPhysics, 2016) and it may be suggested that the nature of the energy is also a precipitation of a range of to be determined meta-functions as suggested in some sample Equations 3.3.2 through 3.3.6. Hence:

$$Gamma\ Rays_{\ Nature\ of\ Energy} = \\ Xa + \overline{Yb_{0-n}}\ \ where \left[\begin{array}{c} X\ \in [S_{System_P}] \\ Y \in [S_{System_{Pr}}, S_{System_P}, S_{System_K}, S_{System_N}] \\ a, b\ are\ integers; a > b \end{array} \right]$$

Eq 3.3.2: Gamma Rays Nature of Energy

$$X - Rays_{Nature\ of\ Energy} = \\ Xa + \overline{Yb_{0-n}}\ \ where \left[\begin{array}{c} X\ \in [S_{System_P}] \\ Y \in [S_{System_{Pr}}, S_{System_{\square}}, S_{System_K}, S_{System_N}] \\ a, b\ are\ integers; a > b \end{array} \right]$$

Eq 3.3.3: X-Rays Nature of Energy

$$Ultraviolet_{\ Nature\ of\ Energy} = \\ Xa + \overline{Yb_{0-n}}\ \ where \left[\begin{array}{c} X\ \in [S_{System_P}] \\ Y \in [S_{System_{Pr}}, S_{System_P}, S_{System_K}, S_{System_N}] \\ a, b\ are\ integers; a > b \end{array} \right]$$

Eq 3.3.4: Ultraviolet Nature of Energy

$$White\ Light_{Nature\ of\ Energy} = Xa + \overline{Yb_{0-n}} \quad where \begin{bmatrix} X \in [S_{System_P}] \\ Y \in [S_{System_{Pr}}, S_{System_P}, S_{System_K}, S_{System_N}] \\ a, b\ are\ integers; a > b \end{bmatrix}$$

Eq 3.3.5: White Light Nature of Energy

$$AM\ Radio\ Waves_{Nature\ of\ Energy} = Xa + \overline{Yb_{0-n}} \quad where \begin{bmatrix} X \in [S_{System_P}] \\ Y \in [S_{System_{Pr}}, S_{System_P}, S_{System_K}, S_{System_N}] \\ a, b\ are\ integers; a > b \end{bmatrix}$$

Eq 3.3.6: AM Radio Waves Nature of Energy

Chapter 3.4: Presence

If there is a large range of frequencies implicit in the electromagnetic spectrum then there is also the possibility of different types of masses implicit in the electromagnetic spectrum. Frequency determines energy, and mass and energy are related through Einstein's famous MC-squared equation. So pushing a little further it is not just that mass and energy are related, but a different kind of frequency or wave-type potentially allows a different type of mass to emerge. So the possibility of different types of mass seems to be related to the property of Presence in Light. In other words the property of Presence emerges as the possibility of different types of masses as suggested by the range of mass-possibilities that can emerge from the electromagnetic spectrum.

Mass can be thought of as a container at U within which all possibility happens. In other words it can be thought of as a precipitation or emergence of system-presence as depicted in Equation 3.4.1:

$$EM\_Spectrum_{Mass\_Possibility} = Xa + \overline{Yb_{0-n}} \quad where \begin{bmatrix} X \in [S_{System_{Pr}}] \\ Y \in [S_{System_{Pr}}, S_{System_{P}}, S_{System_{K}}, S_{System_{N}}] \\ a, b \; are \; integers; a > b \end{bmatrix}$$

Eq 3.4.1: EM Spectrum Mass Possibility

If the frequencies are infinite, then the possibility of the 'types' of masses or matter is also infinite. The mystery of 'Dark Matter' suggested by scientists to be 27% of our universe, as opposed to 5% of visible matter (NASA-darkmatter, 2016) may have some relation to this.

Hypothetically the Equations 3.4.2 through 3.4.6 depict a range of mass possibilities:

$$Gamma\ Rays_{\ Mass\_Possibility} = Xa + \overline{Yb_{0-n}} \ \ where \begin{bmatrix} X \in [S_{System_{Pr}}] \\ Y \in [S_{System_{Pr}}, S_{System_{P}}, S_{System_{K}}, S_{System_{N}}] \\ a, b \ are\ integers; a > b \end{bmatrix}$$

Eq 3.4.2: Gamma Rays Mass Possibility

$$Ultraviolet_{\ Mass\_Possibility} = Xa + \overline{Yb_{0-n}} \ \ where \begin{bmatrix} X \in [S_{System_{Pr}}] \\ Y \in [S_{System_{Pr}}, S_{System_{P}}, S_{System_{K}}, S_{System_{N}}] \\ a, b \ are\ integers; a > b \end{bmatrix}$$

Eq 3.4.3: Ultraviolet Mass Possibility

$$Blue\ Light_{\ Mass\_Possibility} = Xa + \overline{Yb_{0-n}} \ \ where \begin{bmatrix} X \in [S_{System_{Pr}}] \\ Y \in [S_{System_{Pr}}, S_{System_{P}}, S_{System_{K}}, S_{System_{N}}] \\ a, b \ are\ integers; a > b \end{bmatrix}$$

Eq 3.4.4: Blue Light Mass Possibility

$$Microwaves_{\ Mass\_Possibility} = Xa + \overline{Yb_{0-n}} \ \ where \begin{bmatrix} X \in [S_{System_{Pr}}] \\ Y \in [S_{System_{Pr}}, S_{System_{P}}, S_{System_{K}}, S_{System_{N}}] \\ a, b \ are\ integers; a > b \end{bmatrix}$$

Eq 3.4.5: Microwaves Mass Possibility

$$FM\ Radio\ Waves_{Mass\_Possibility} = Xa + \overline{Yb_{0-n}} \quad where \begin{bmatrix} X \in [S_{System_{Pr}}] \\ Y \in [S_{System_{Pr}}, S_{System_P}, S_{System_K}, S_{System_N}] \\ a, b\ are\ integers;\ a > b \end{bmatrix}$$

Eq 3.4.6: FM Radio Waves Mass Possibility

SECTION 4: EMERGENCE OF MATTER

So we find that the four underlying properties of Light that we call Harmony, Knowledge, Power, and Presence are of the essence of the speed with which the electromagnetic spectrum moves, the wave-range within the electromagnetic spectrum, the energy-gradient within the electromagnetic spectrum, and the mass-possibilities due to the electromagnetic spectrum, respectively. Further the description – electromagnetic – seems to have captured the Power-Harmony aspects implicit in light. In reality the electro-magnetic spectrum can likely be more completely described as wavearchetype-electro-magnetic-masspotential spectrum.

But further, we will find that layers of matter – quantum particles, which include bosons, and atoms – are also structured or emerge along the same property-lines or property-families of light.

Chapter 4.1: Quantum particles

The electromagnetic spectrum or perhaps more accurately, the wavearchetype-electro-magnetic-masspotential spectrum, describes one of the first layers or translation of light into manifestation. This spectrum or field is an example of what stands behind visible matter and perhaps it is fair to say that all matter emerges from such spectrums.

As already explored the finite speed of such spectrums allows the build-up of energies or quanta, and this in turn may express itself as a series or as an array of quantum particles. The vast number of quantum particles that have so far been discovered has in turn yielded what is known as the Standard Model. This model is made up of what is called Quarks, Leptons, and Bosons, and a Higgs-Boson (Cottingham, 2007).

But if we look at these four fundamental quantum categories of particles from a property-based, or "functional" viewpoint a different chapter in the exploiration of light emerges.

Quarks

For instance, it can be seen first that the nucleus of an atom is made up of a combination of quarks. Specifically, a proton is composed of two "up" quarks and one "down" quark. Quarks have unusual names – up, down, charm, strange, top, bottom, with each subsequent pair belonging to a different "generation" of quarks. A neutron is composed of two "down" quarks and one "up" quark. Protons and neutrons together make up the nucleus of an atom. But we also know that the number of protons in the nucleus specifies the Atomic Number of an atom. Atomic number in turn uniquely identifies the element from the periodic table.

Hence, an atomic number of 47, for example, specifies that the element is Silver. In other words it can be suggested that the unique properties of an element, the knowledge of what it is and how it will behave in the universe, is related to the quark. It may be suggested that quarks, therefore, are associated with the precipitation or emergence of Light's property of Knowledge in the quantum world.

Hence, it could be that the signature for the family of quarks, as in Equation 4.1.1, is:

$$Sig_{quarks} = \frac{Xa + }{Yb_{0-n}} \quad where \begin{bmatrix} X \in [S_{System_K}] \\ Y \in [S_{System_{Pr}}, S_{System_P}, S_{System_K}, S_{System_N}] \\ a, b \text{ are integers}; a > b \end{bmatrix}$$

Eq 4.1.1: Generalized Signature of Quarks

As can be seen the primary element X is derived from the set of knowledge, S_{System_K}. Various elements, derived from the four sets would define the behavior of quarks and could be functions such as 'composite-arrangements', 'specifying attributes' amongst others, hence collectively describing quarks' way of being.

Leptons

When considering leptons it is useful to know that unlike quarks that only exist in composite arrangements with other quarks, leptons are solitary, point-like particles without internal structure (Arabatzis, 2006). The best-known lepton is the electron. So the electron may be considered as a surrogate for the lepton class. The electron appears to be associated with the flow of energy and power. Further they appear to be the adventurers easily leaving the atom they are a part of. They also form locks or bonds with other atoms through the force of attraction and repulsion. In some sense they seem to be a representation or precipitation or emergence of Light's property of Power.

The signature for the family of leptons, as in Equation 4.1.2, is:

$$Sig_{leptons} = Xa + \overline{Yb_{0-n}} \quad where \begin{bmatrix} X \in [S_{System_P}] \\ Y \in [S_{System_{Pr}}, S_{System_P}, S_{System_K}, S_{System_N}] \\ a, b \text{ are integers}; a > b \end{bmatrix}$$

Eq 4.1.2: Generalized Signature of Leptons

As can be seen the primary element X is derived from the set of power, S_{System_P}. Various elements, derived from the four sets would define the behavior of leptons and could be functions such as 'adventurer', 'solitary', 'create attraction', amongst others, hence collectively describing leptons' way of being.

Bosons

The bosons on the other hand are thought of as force-carriers. They are what allow all known matter particles to interact. The three fundamental bosons in this category are the photon, the W and Z bosons, and the gluon. The carrier particle of the electromagnetic force or spectrum is the photon. The carrier particle of the strong nuclear force that holds quarks together is the gluon. The carrier particle for the weak interactions, responsible for the decay of massive quarks and leptons into lighter quarks and leptons, are the W and Z bosons.

Bosons can be thought of as the precipitation of what creates relationship and harmony at the quantum level. Hence they can be thought of as the precipitation or emergence of Light's property of Harmony at the quantum level.

The signature for the family of gauge bosons, as in Equation 4.1.3, is:

$$Sig_{bosons} = \overline{\begin{matrix} Xa + \\ Yb_{0-n} \end{matrix}} \quad where \begin{bmatrix} X \in [S_{System_N}] \\ Y \in [S_{System_{Pr}}, S_{System_P}, S_{System_K}, S_{System_N}] \\ a, b \text{ are integers}; a > b \end{bmatrix}$$

Eq 4.1.3: Generalized Signature of Bosons

As can be seen the primary element X is derived from the set of nurturing, S_{System_N}. Various elements, derived from the four sets would define the behavior of bosons and could be functions such as 'creating interaction', 'holding together', amongst others, hence collectively describing bosons' way of being.

Equations in this form give a clue as to how to think about the fundamental particles. While the default approach is always to think about physical qualities, equations in the preceding format allow us to think about detailed functions or qualities associated with particles and give insight into how the process of emergence or complexification of four-fold functionality takes place.

Higgs-Boson

This leaves the other discovered fundamental particle the Higgs-Boson. In ordinary matter, most of the mass is contained in atoms, and the majority of the mass of an atom resides in the nucleus, made of protons and neutrons. Protons and neutrons are each made of three quarks. But it is the quarks that get their mass by interacting with what is known as the Higgs field (Olive,

2014). Hence the Higgs-Boson can be thought of as the mass-giver. In other words it is what gives presence to the quarks and it can be thought of as the precipitation or emergence of Light's property of Presence at the quantum level. Just as there are multiple particles in each of the other 'families' it is likely that there will be multiple particles in the Higgs-Boson family. Recent research at CERN indicates that the Higgs-Boson may have a cousin.

The signature for the Higgs-boson and any other similar particle, as in Equation 4.1.4, is:

$$Sig_{Higgs-boson} = Xa + \overline{Yb_{0-n}} \quad where \begin{bmatrix} X \in [S_{System_{Pr}}] \\ Y \in [S_{System_{Pr}}, S_{System_{P}}, S_{System_{K}}, S_{System_{N}}] \\ a, b \text{ are integers}; a > b \end{bmatrix}$$

Eq 4.1.4: Generalized Signature of Higgs-Boson

As can be seen the primary element X is derived from the set of presence, $S_{System_{Pr}}$. Various elements, derived from the four sets would define the behavior of Higgs-bosons and could be a function such as 'creating mass', amongst others, hence collectively describing Higgs-bosons' way of being.

Combining all the preceding particle equations it is possible to create a generalized particle equation, as in Equation 4.1.5:

$$Sig_{particle} = Xa + \overline{Yb_{0-n}} \quad where \begin{bmatrix} X \in [S_{System_{Pr}}, S_{System_{P}}, S_{System_{K}}, S_{System_{N}}] \\ Y \in [S_{System_{Pr}}, S_{System_{P}}, S_{System_{K}}, S_{System_{N}}] \\ a, b \text{ are integer}\square; a > b \end{bmatrix}$$

Eq 4.1.5: Generalized Signature of Particle

So we see that again the underlying properties of light - Knowledge, Power, Harmony, and Presence - emerge as quarks, leptons, bosons, and Higgs-bosons respectively.

Chapter 4.2: Bosons

The bosons as mentioned can be thought of as force-carriers and allow all known matter particles to interact.

But when we look at bosons in more detail there are three fundamental bosons – the photon, the W and Z bosons, the gluon - and one hypothetical boson – the graviton.

Photon

The photon is the carrier particle of the electromagnetic force. But in the scheme of things the electromagnetic force pervades everything and as explored in the section on the electromagnetic spectrum appears to be foundational to this reality we are in. So it could be said that it is related to Presence or is an emergence of Light's property of Presence at the level of quantum force-carriers.

Hence, an equation for the photon, Equation 4.2.1 could be:

$$Sig_{photon} = Xa + \overline{Yb_{0-n}} \quad where \begin{bmatrix} X \in [S_{System_N}] \\ Y \in [S_{System_{PT}}, S_{System_P}, S_{System_K}, S_{System_N}] \\ a, b \text{ are integers}; a > b \end{bmatrix}$$

Eq 4.2.1: Signature of Photon

While the primary element X, would continue to be the same as X for bosons, as in Equation 4.1.3, there will be additional secondary elements Y that will further qualify the attributes or functionality of photons. As an example such elements could be 'pervasiveness', 'multiple wave handler', amongst others.

Gluon

The gluon is the carrier particle of what is known as the strong nuclear force and holds quarks together in their inherently composite arrangements. But we have posited that quarks are related to or are an emergence of Knowledge at the quantum-particle level. Hence it must be that the gluon is an emergence of Light's property of Knowledge at the level of quantum force-carriers.

Hence, Equation 4.2.2 is an equation for the gluon:

$$Sig_{gluon} = \frac{Xa + }{Yb_{0-n}} \quad where \begin{bmatrix} X \in [S_{System_N}] \\ Y \in [S_{System_{Pr}}, S_{System_P}, S_{System_K}, S_{System_N}] \\ a, b \text{ are integers}; a > b \end{bmatrix}$$

Eq 4.2.2: Signature of Gluon

The primary element X would continue to be the same as in the equation for the boson. There are likely additional secondary elements Y such as 'concentrated', 'intense connection', amongst others that would collectively determine the behavior of gluons.

W and Z Bosons

The W and Z bosons are the carrier particle for the weak interactions, responsible for the decay of massive quarks and leptons into lighter quarks and leptons. But this usually is accompanied by the release of energy and power and so it must be that W and Z bosons are an emergence of Light's property of Power at the level of quantum force-carriers.

Hence, Equation 4.2.3, for W and Z bosons is as follows:

$$Sig_{W,Z\ bosons} = Xa + \overline{Yb_{0-n}}\ \ where \begin{bmatrix} X \in [S_{System_N}] \\ Y \in [S_{System_{Pr}}, S_{System_P}, S_{System_K}, S_{System_N}] \\ a, b\ are\ integers; a > b \end{bmatrix}$$

Eq 4.2.2: Signature of W and Z Bosons

The primary element X would continue to be the same as in the equation for the boson. There are likely additional secondary elements Y such as 'weak link', 'release of energy', amongst others that would collectively determine the behavior of W and Z bosons.

Graviton

This leaves the hypothetical graviton that is thought to be the carrier particle for the force of gravity. But gravity is what holds astronomical objects together in relationship. Hence it must be that gravitons are an emergence of Light's property of Harmony at the level of quantum force-carriers.

Hence, Equation 4.2.4, for the hypothetical graviton is as follows:

$$Sig_{graviton} = Xa + \overline{Yb_{0-n}}\ \ where \begin{bmatrix} X \in [S_{System_N}] \\ Y \in [S_{System_{Pr}}, S_{System_P}, S_{System_K}, S_{System_N}] \\ a, b\ are\ integers; a > b \end{bmatrix}$$

Eq 4.2.2: Signature of W and Z Bosons

The primary element X would continue to be the same as in the equation for the boson. There are likely additional secondary elements Y such as 'mass curving space', 'space defining object movement', amongst others that would collectively determine the behavior of gravitons.

So we see that the underlying properties of light - Presence, Knowledge, Power, and Harmony - emerge as photons, gluons, W and Z bosons, and the hypothetical graviton at the quantum force-carrier level.

Chapter 4.3: Atoms

Quantum particles in turn create atoms, and all known atoms can be classified into one of four groups: The s-group, p-group, d-group, and f-group. But what are these groups and are they too emergences of Light as it continues its journey of crystalizing the possibility or potentiality or fourfold functionality within it?

s-Group

The s-group consists primarily of what are known as alkali metals and alkali earth metals. These alkali metals are known to easily lose electrons and form what is known as positive ions. When they lose electrons energy is gained, but when the electrons are taken up by other atoms in proximity there is a lot of energy released. Some have referred to these groups as "violent worlds" (Tweed, 2003), and it has been pointed out that stars shine because they are transmuting vast amount of hydrogen into helium, both of which are s-group elements. So one gets the sense that the s-group may be an emergence of Light's property of Power.

Stars and suns are also known to be the crucibles where all the different kinds of atoms are created. So these furnaces of power by virtue of their heat and high pressure are able to force electrons and protons and neutrons to come together to create all the different types of atoms known in the universe.

But philosophically what are the s-group atoms or elements? These are atoms where there is an equal likelihood of an electron being anywhere in a symmetrical sphere around the nucleus. The other groups are all similarly defined by likelihoods of electrons being within a possible pattern around the nucleus. The patterns that distinguishes s-group atoms is a sphere, and since all other patterns can be thought of as occurring within some sphere, in some sense this is like an imprint or precipitation or emergence that allows other kinds of emergences to surface within it. So the elements that are part of the s-group may be thought of as the adventurers with courage who venture into a brave new world to create some foundation by which all other element-creations can follow.

The fact that hydrogen and helium are known to constitute 98% of the Universe (Heiserman, 1991) relative to other elements therefore makes sense in this view, especially since hydrogen and helium provide the fuel with which the star-furnaces manufacture all other elements.

So s-group elements seem to embody functions such as power, energy, adventure, courage, and can be thought of as an emergence of the property of Light to do with Power.

Hence, a series of equations linked to S_{System_P} as the prime set can be suggested, starting with the s-group mapping, as in Equation 4.3.1:

$$Element_{s-group} = Xa + \overline{Yb_{0-n}} \quad where \begin{bmatrix} X \in [S_{System_P}] \\ Y \in [S_{System_{Pr}}, S_{System_P}, S_{System_K}, S_{System_N}] \\ a, b \text{ are integers}; a > b \end{bmatrix}$$

Eq 4.3.1: s-Group Element

Further, the equivalent mapping between traditional element groupings and S_{System_P} as in Equations 4.3.2 and 4.3.3 can be specified:

$$Element_{\ Alkali\ metal} = Xa + \overline{Yb_{0-n}}\ \ where \left[\begin{matrix} X\ \in [S_{Systemp}] \\ Y \in [S_{Systemp_{Pr}}, S_{Systemp}, S_{Systemp_K}, S_{Systemp_N}] \\ a, b\ are\ integers; a > b \end{matrix}\right]$$

Eq 4.3.2: Alkali Metal Element

$$Element_{\ Alkali\ earth\ metal} = Xa + \overline{Yb_{0-n}}\ \ where \left[\begin{matrix} X\ \in [S_{Systemp}] \\ Y \in [S_{Sy\square temp_{Pr}}, S_{Systemp}, S_{Systemp_K}, S_{Systemp_N}] \\ a, b\ are\ integers; a > b \end{matrix}\right]$$

Eq 4.3.3: Alkali Earth Metal Element

Further, as a representative element belonging to the Alkali Metal group the equation for Lithium (Li), as in Equation 4.3.4, would be:

$$Element_{\ Lithium} = Xa + \overline{Yb_{0-n}}\ \ where \left[\begin{matrix} X\ \in [S_{Systemp}] \\ Y \in [S_{Systemp_{Pr}}, S_{Systemp}, S_{Systemp_K}, S_{Systemp_N}] \\ a, b\ are\ integers; a > b \end{matrix}\right]$$

Eq 4.3.4: Lithium

Note that the primary element X in all these cases may be a function or attribute along the lines of 'expresses power'. The secondary elements Y in all these would have some elements being the same, and then would have specific differences as one gets into sub-classes and the individual elements themselves.

p-Group

Atoms or elements belonging to the p-group are those with the likelihood of electrons occurring equally on either side of the nucleus, like a dumbbell of sorts.

There are some very significant elements in this group that are part of the metal, metalloid, non-metal, halogen, and noble gas sub-groupings. Carbon, Nitrogen, Oxygen, and Silicon are some of the sample elements. Looking at the types of elements present in this group it is as though all the element possibilities have been represented within it. It is perhaps that the possibility of ideas behind all elements has emerged in this group and one can hypothesize that this group may be a reflection of the property of Knowledge, forming archetypes from which all other elements are created.

Philosophically, the one spherical shell or probability cloud (s) becoming two shells like a dumbbell (p) signifies the creation of an essential polarity within a unit space. So in a sense the one becoming two is the first instance of variability in space. The two, spaced in 3-dimensions around the nucleus in a sense create six switches becoming an attractor or allowing a vaster number of different kinds of elements to surface. So there is a sense of the 'idea' of the element that comes into focus.

But further, the essential elements that allow both thinking and virtual thinking machines to come into being, are also contained within this group. Carbon is the basis of DNA and of all life. The fact the Silicon, directly below it in the periodic table and therefore sharing essential qualities, is considered the basis of all virtual thinking machines is therefore perhaps significant and may reinforce the notion that the p-group is a precipitation of Light's property of Knowledge.

Hence, a series of equations linked to S_{System_K} as the prime set can be suggested, starting with the p-group mapping, as in Equation 4.3.5:

$$Element_{\,p-group} = Xa + \overline{Yb_{0-n}} \;\; where \begin{bmatrix} X \in [S_{System_K}] \\ Y \in [S_{System_{Pr}}, S_{System_P}, S_{System_K}, S_{System_N}] \\ a, b \; are \; integers; a > b \end{bmatrix}$$

Eq 4.3.5: p-group Element

Further, the equivalent mapping between traditional element groupings and S_{System_K} as in Equations 4.3.6 through 4.3.11 can also be specified:

$$Element_{\,Metal} = Xa + \overline{Yb_{0-n}} \;\; where \begin{bmatrix} X \in [S_{System_K}] \\ Y \in [S_{System_{Pr}}, S_{System_P}, S_{System_K}, S_{System_N}] \\ a, b \; are \; integers; a > b \end{bmatrix}$$

Eq 4.3.6: Metal Element

$$Element_{\,Metalloid} = Xa + \overline{Yb_{0-n}} \;\; where \begin{bmatrix} X \in [S_{System_K}] \\ Y \in [S_{System_{Pr}}, S_{System_P}, S_{System_K}, S_{System_N}] \\ a, b \; are \; integers; a > b \end{bmatrix}$$

Eq 4.3.7: Metalloid Element

$$Element_{\,Non-Metal} = Xa + \overline{Yb_{0-n}} \;\; where \begin{bmatrix} X \in [S_{System_K}] \\ Y \in [S_{System_{Pr}}, S_{System_P}, S_{System_K}, S_{System_N}] \\ a, b \; are \; integers; a > b \end{bmatrix}$$

Eq 4.3.8: Non-Metal Element

$$Element_{Halogen} = Xa + \overline{Yb_{0-n}} \quad where \begin{bmatrix} X \in [S_{System_K}] \\ Y \in [S_{System_{Pr}}, S_{System_P}, S_{System_K}, S_{System_N}] \\ a, b \text{ are integers}; a > b \end{bmatrix}$$

Eq 4.3.9: Halogen Element

$$Element_{Noble\ Gas} = Xa + \overline{Yb_{0-n}} \quad where \begin{bmatrix} X \in [S_{System_K}] \\ Y \in [S_{System_{Pr}}, S_{System_P}, S_{System_K}, S_{System_N}] \\ a, b \text{ are integers}; a > b \end{bmatrix}$$

Eq 4.3.10: Noble Gas Element

Further, as a representative element belonging to the Non-Metal group the equation for Carbon (C), as in Equation 4.3.11, would be:

$$Element_{Carbon} = Xa + \overline{Yb_{0-\square}} \quad where \begin{bmatrix} X \in [S_{System_K}] \\ Y \in [S_{System_{Pr}}, S_{System_P}, S_{System_K}, S_{System_N}] \\ a, b \text{ are integers}; a > b \end{bmatrix}$$

Eq 4.3.11: Carbon

Note that the primary element X in all these cases may be a function or attribute along the lines of 'knowledge'. The secondary elements Y in all these would have some elements being the same, and then would have specific differences as one gets into sub-classes and the individual elements themselves.

d-Group

The d-group comprises the Transition Metals. These metals are generally hard and strong, exhibit corrosive resistance, and can be thought of as workhorse elements. Many industrial and well-known elements sit in this group: Titanium, Chromium, Manganese, Iron, Cobalt, Nickel, Copper, Zinc, Silver, Platinum, and Gold, amongst others.

The elements and atoms in the d-group are equally likely to show up in four possible lobes or probability-spaces around the nucleus. Four lobes occurring in multiple possible planes around the nucleus will likely create a space of stability, since there is a possibility of four lobes creating the four vertices of a tetrahedron (Fuller, 1982) that has been positioned as one of the most stable shapes in the universe.

Much of the constructed world around us is created from these elements. Further, most of the series in the group easily lose one or more electrons thereby easily combining with other atoms to form a vast array of compounds. Also, looking more broadly at the function of these elements, it can be seen that these metals exist for service, to help bring about perfection in the constructed world, to help much of the machinery in which they are used, and to assist the processes dependent on them to be completed with diligence. Hence, these transition metals appear to be an emergence of Light's property of Presence.

Therefore, a series of equations linked to $S_{System_{Pr}}$ as the prime set can be suggested, starting with the d-group mapping, as in Equation 4.3.12:

$$Element_{d-group} = Xa + \overline{Yb_{0-n}} \quad where \begin{bmatrix} X \in [S_{System_{Pr}}] \\ Y \in [S_{System_{Pr}}, S_{System_{P}}, S_{System_{K}}, S_{System_{N}}] \\ a, b \text{ are integers}; a > b \end{bmatrix}$$

Eq 4.3.12: d-group Element

Further, the equivalent mapping between traditional element groupings and $S_{System_{Pr}}$, as in Equation 4.3.13, can also be specified:

$$Elem\square nt_{Transition\ Metal} = Xa + \overline{Yb_{0-n}} \quad where \begin{bmatrix} X \in [S_{System_{Pr}}] \\ Y \in [S_{System_{Pr}}, S_{System_{P}}, S_{System_{K}}, S_{System_{N}}] \\ a, b \text{ are integers}; a > b \end{bmatrix}$$

Eq 4.3.13: Transition Metal Element

Further, as a representative element belonging to the Transition-Metal group the equation for Gold (Au), as in Equation 4.3.14, would be:

$$Element_{Gold} = Xa + \overline{Yb_{0-n}} \quad where \begin{bmatrix} X \in [S_{System_{Pr}}] \\ Y \in [S_{System_{Pr}}, S_{System_{P}}, S_{System_{K}}, S_{System_{N}}] \\ a, b \text{ are integers}; a > b \end{bmatrix}$$

Eq 4.3.14: Gold

Note that the primary element X in all these cases may be a function or attribute along the lines of 'work horse'. The secondary elements Y in all these would have some elements being the same, and then would have specific differences as one gets into sub-classes and the individual elements themselves.

f-Group

The f-group comprises of the Lanthanides and Actinides. Philosophically, elements in the f-group consist of 6 lobes around the nucleus within which an electron may be found. 6 lobes will exist in multiple planes around the nucleus and suggests the notion of extended relationship and collectivity: the attempt to build larger and larger bonds within a small space. Considering this it is likely that the f-group is an emergence of Light's property of Harmony.

Thinking about Lanthanides, some interesting facts may reinforce this notion. First, the spin of electrons in a lanthanides' outer shell is aligned, creating a strong magnetic field. The notion of creating a strong magnetic field seems to be consistent with the notion of engendering a collectivity through the ordered attraction and repulsion of elements. Second, these elements curiously occur together in nature often in the same ores (Gray, 2009) and are chemically interchangeable also suggesting the notion of forming a tight intra-group collectivity.

Actinides on the other hand are inherently radioactive. This implies that these elements have inherently crossed a threshold of stability and have the urge, over their own half-lives, to decompose into other elements. This natural urge may suggest some boundary conditions on the notion of collectivity and nurturing, giving additional insight into the nature of collectivity and nurturing. Further, the entire actinide group, as opposed to the lanthanide group that is inherently stable, is unstable. It is curious that both these should be part of the f-group, and they must provide insight into boundary conditions into the notion of collectivity in elements.

Hence, a series of equations linked to S_{System_N} as the prime set can be suggested, starting with the f-orbital mapping, as in Equation 4.3.15:

$$Element_{f-group} = Xa + \overline{Yb_{0-n}} \; where \begin{bmatrix} X \in [S_{System_N}] \\ Y \in [S_{System_{Pr}}, S_{System_P}, S_{System_K}, S_{System_N}] \\ a, b \; are \; integers; a > b \end{bmatrix}$$

Eq 4.3.15: f-group Element

Further, the equivalent mapping between traditional element groupings and S_{System_N}, as in Equations 4.3.16 and 4.3.17, can also be specified:

$$Element_{Lanthanide} = Xa + \overline{Yb_{0-n}} \; where \begin{bmatrix} X \in [S_{System_N}] \\ Y \in [S_{System_{Pr}}, S_{System_P}, S_{System_K}, S_{System_N}] \\ a, b \; are \; integers; a > b \end{bmatrix}$$

Eq 4.3.16: Lanthanide Element

$$Element_{Actinide} = Xa + \overline{Yb_{0-n}} \; where \begin{bmatrix} X \in [S_{System_N}] \\ Y \in [S_{System_{Pr}}, S_{System_P}, S_{System_K}, S_{System_N}] \\ a, b \; are \; integers; a > b \end{bmatrix}$$

Eq 4.3.17: Actinide Element

Further, as a representative element belonging to the Lanthanide group the equation for Lanthanum (La), as in Equation 4.3.18, would be:

$$Element_{Lanthanum} = Xa + \overline{Yb_{0-n}} \quad where \begin{bmatrix} X \in [S_{System_N}] \\ Y \in [S_{System_{Pr}}, S_{System_P}, S_{System_K}, S_{System_N}] \\ a, b\ are\ integers; a > b \end{bmatrix}$$

Eq 4.3.18: Lanthanum

Note that the primary element X in all these cases may be a function or attribute along the lines of 'experiments in collectivity'. The secondary elements Y in all these would have some elements being the same, and then would have specific differences as one gets into sub-classes and the individual elements themselves.

SECTION 5: EMERGENCE OF LIFE

This Section explores the emergence of Life through the Cell, complex human attributes such as thoughts and feelings, and uniqueness of Nations.

Hence Chapter 5.1 will explore the emergence and complexification of four-foldness through the primary molecular plans of nucleic acids, proteins, lipids, and polysaccharides and position the various combinations of this four-foldness as the basis of Natural History.

Chapter 5.2 will relate key human attributes of sensations, urges, desires, wills, feelings, emotions, and thought to the continuing journey of fourfold complexification.

Chapter 5.3 will suggest how every country is itself a manifestation, in essence, of various combinations of Light's implicit fourfold properties.

Chapter 5.1: Living Cells

The living cell has a universe of adaptability embedded in it. In 'The Machinery of Life', Goodsell, an Associate Professor of Molecular Biology at the Scripps Research Institute (Goodsell, 2010) suggests that every living thing on Earth uses a similar set of molecules to eat, to breathe, to move, and to reproduce. There are molecular machines that do the myriad things that distinguish living organisms that are identical in all living cells. This nanoscale machinery of cells uses four basic molecular plans with unique chemical personalities: nucleic acids, proteins, lipids, and polysaccharides.

But further, it is a combination of these molecular plans that can cause specialization of a cell to create any feature of a plant or animal. At the level of plants for example, striking colors and scents can become the natural features of a plant. It can adapt its shape to allow various insects or bees to enter into it in different ways. It can store a myriad of nutrients in a myriad different ways. It can develop poisons to keep predatory animals away. At the level of animals for example, skins can toughen into hard shells. Feet can develop suction caps to allow mobility in a vertical direction. Eyes can develop broad vision, telescopic vision, or microscopic vision. The mechanics of all adaptability can be traced to the ability of the four molecular plans to coordinate and generate more granular cell-level features that give it some evolutionary advantage in the scheme of things. All Natural History, hence, can be traced to this ability of cells to specialize along a diverse array of specialized lines.

In this chapter we will examine the mathematics of the molecular plans that form the bases for such specialization to occur.

Nucleic Acids

Nucleic acids basically encode information. They store and transmit the genome, the hereditary information needed to keep the cell alive. They function as the cell's librarians and contain information on how to make proteins and when to make them.

They are hence, the keepers of a cell's knowledge, its wisdom, its ability to make laws, the vehicle to spread knowledge within cells and to the next generation of cells. Being so, one can see that there is similarity with the set for system-knowledge highlighted earlier in Chapter 2.3. Reproducing Equation 2.3.3:

$$S_{System_K} \ni [Wisdom, Law\ Making, Spread\ of\ Knowledge\ ...]$$

Nucleic acids can therefore be thought of as a precipitation of system-knowledge at the cellular level.

Hence, a nucleic acid will have a generalized signature, as in Equation 5.1.1, derived from the system-knowledge family:

$$Sig_{nucleic\ acid} = Xa + \overline{Yb_{0-n}} \quad where \begin{bmatrix} X \in [S_{System_K}] \\ Y \in [S_{System_{Pr}}, S_{System_P}, S_{System_K}, S_{System_N}] \\ a, b\ are\ integers; a > b \end{bmatrix}$$

Eq 5.1.1: Nucleic Acid

The primary element X could be an attribute or function such as 'keeper of knowledge'. Secondary elements Y could be 'protein laws', 'generational knowledge', amongst others. The collectivity of elements as per the equation would specify the character of a nucleic acid.

DNA and RNA, two types of nucleic acids, would hence have the equations as specified by Equation 5.1.2 and 5.1.3 respectively.

$$Sig_{DNA} = Xa + \overline{Yb_{0-n}} \quad where \begin{bmatrix} X \in [S_{System_K}] \\ Y \in [S_{System_{P_T}}, S_{System_P}, S_{System_K}, S_{System_N}] \\ a, b\ are\ integers; a > b \end{bmatrix}$$

Eq 5.1.2: DNA

$$Sig_{RNA} = Xa + \overline{Yb_{0-n}} \quad where \begin{bmatrix} X \in [S_{System_K}] \\ Y \in [S_{System_{P_T}}, S_{System_P}, S_{System_K}, S_{System_N}] \\ a, b\ are\ integers; a > b \end{bmatrix}$$

Eq 5.1.3: RNA

The primary element X in (5.1.2) and (5.1.3) would be the same as that for nucleic acids (5.1.1). The secondary elements Y however will be a larger and more specific set with many elements in common with (5.1.1).

Proteins

Proteins are the cells work-horses. Look anywhere in a cell and one will see proteins at work. Proteins are built in thousands of shapes and sizes, each performing a different function. As Goodsell describes, "some are built simply to adopt a defined shape, assembling into rods, nets, hollow spheres, and tubes. Some are molecular motors, using energy to rotate, or flex, or crawl. Many are chemical catalysts that perform chemical reactions atom-by-atom, transferring and transforming chemical groups exactly as needed." With their wide potential for diversity, proteins are constructed to perform most of the everyday tasks of the cells. In fact human cells build around 30,000 different kinds of proteins to execute on the diverse array of cellular level tasks.

Proteins hence, exist for service, to bring about perfection at the level of the cell, are characterized by extreme diligence and perseverance, and so on. Being so, one can see that there is similarity with the set for system-presence highlighted earlier in Chapter 2.3. Reproducing Equation 2.3.1:

$$S_{System_{Pr}} \ni [Service, Perfection, Diligence, Perseverance, \ldots]$$

Proteins can therefore be thought of as a precipitation of system-presence at the cellular level.

Hence, a protein could have a generalized signature, as in Equation 5.1.4, derived from the system-presence family:

$$Sig_{protein} = Xa + \overline{Yb_{0-n}} \quad where \left[\begin{array}{c} X \in [S_{System_{Pr}}] \\ Y \in [S_{System_{Pr}}, S_{System_{P}}, S_{System_{K}}, S_{System_{N}}] \\ a, b \; are \; integers; a > b \end{array}\right]$$

Eq 5.1.4: Protein

This could yield a vast number of functional proteins. In fact it may be possible that the 30,000 or so known proteins created by the human cell could each be specified by a signature equation of this nature. It may be possible to map existing proteins to functionality as suggested by the four sets of molecular plans.

Consider Insulin, for example. Insulin regulates the metabolism of carbohydrates, fats and protein by promoting the absorption of, especially, glucose from the blood into fat, liver and skeletal muscle cells. Equation 5.1.5 for Insulin would hence be:

$$Sig_{insulin} = Xa + \overline{Yb_{0-n}} \quad where \begin{bmatrix} X \in [S_{System_{Pr}}] \\ Y \in [S_{System_{Pr}}, S_{System_{P}}, S_{System_{K}}, S_{System_{N}}] \\ a, b \text{ are integers}; a > b \end{bmatrix}$$

Eq 5.1.5: Insulin

The primary element X could be an attribute or function such as 'workhorse. Secondary elements Y could be 'metabolic regulation, 'glucose absorption', 'blood to fat channel', amongst others. The collectivity of elements as per the equation would specify the character of insulin.

Consider Histones as another example. They are the chief protein components of chromatin, acting as spools around which DNA winds, and playing a role in gene regulation. Without histones, the unwound DNA in chromosomes would be very long (a length to width ratio of more than 10 million to 1 in human DNA). Equation 5.1.6 for Histones would hence be:

$$Sig_{histones} = Xa + \overline{Yb_{0-n}} \quad where \begin{bmatrix} X \in [S_{System_{Pr}}] \\ Y \in [S_{System_{Pr}}, S_{Sy\square temp}, S_{System_{K}}, S_{System_{N}}] \\ a, b \text{ are integers}; a > b \end{bmatrix}$$

Eq 5.1.6: Histones

The secondary element Y would have elements such as 'gene regulation', 'unwound DNA management', amongst others.

Lipids

Lipids by themselves are tiny molecules, but when grouped together form the largest structures of the cell. When placed in water, lipid molecules aggregate to form huge waterproof sheets. These sheets easily form boundaries at multiple levels and allow concentrated interactions and work to be performed within a cell. Hence, the nucleus and the mitochondria are contained within lipid-defined compartments. Similarly, each cell itself is contained within a lipid-defined boundary.

Lipids are therefore promoters of relationship, of harmony in the cell, of nurturing the cell-level division of labor, of allowing specialization and uniqueness to emerge, hence perhaps of earlier forms of compassion and love, and so on. The notion of such early forms of compassion is consistent with the biologist's perspective that at some point a gene for compassion was developed in pre-human species (Wright, 2009). Being so, one can see that there is similarity with the set for system-nurturing highlighted earlier in Chapter 2.3. Reproducing Equation 2.3.4:

$$S_{System_N} \ni [Love, Compassion, Harmony, Relationship\ \ldots]$$

This function of harmonization suggests that lipids can therefore be thought of as a precipitation of system-nurturing at the cellular level.

Lipids could have a generalized signature, as in Equation 5.1.7, derived from the system-nurturing family:

$$Sig_{lipid} = Xa + \overline{Yb_{0-n}} \quad where \begin{bmatrix} X \in [S_{System_N}] \\ Y \in [S_{System_{Pr}}, S_{System_P}, S_{System_K}, S_{System_N}] \\ a, b\ are\ integers; a > b \end{bmatrix}$$

Eq 5.1.7: Lipid

Specific lipids such as monoglycerides and phospholipids cold have the following equations:

$$Sig_{monoglyceride} = Xa + \overline{Yb_{0-n}} \quad where \begin{bmatrix} X \in [S_{System_N}] \\ Y \in [S_{System_{Pr}}, S_{System_P}, S_{System_K}, S_{System_N}] \\ a, b\ are\ integers; a > b \end{bmatrix}$$

Eq 5.1.8: Monoglyceride

$$Sig_{phospholipids} = Xa + \overline{Yb_{0-n}} \quad where \begin{bmatrix} X \in [S_{System_N}] \\ Y \in [S_{System_{Pr}}, S_{System_P}, S_{System_K}, S_{System_N}] \\ a, b\ are\ integers; a > b \end{bmatrix}$$

Eq 5.1.9: Phospholipids

The primary element X shared by each of the lipids could be an attribute or function such as 'compartmentalization'. Secondary elements Y could be of the nature of 'work breakdown', 'intra-cell love', amongst others, and would vary with each different kind of lipid.

Polysaccharides

Polysaccharides are long, often branched chains of sugar molecules. Sugars are covered with hydroxyl groups, which associate to form storage containers. As a result polysaccharides function as the storehouse of cell's energy. In addition polysaccharides are also used to build some of the most durable biological structures. The stiff shell of insects, for example are made of long polysaccharides.

Polysaccharides function to create energy, power, courage, strength thereby readying the cell for adventure, and so on. Being so, one can see that there is similarity with the set for system-power highlighted previously in Chapter 2.3. Reproducing Equation 2.3.2:

$$S_{System_P} \ni [Power, Courage, Adventure, Justice, \dots]$$

Providing energy and strength, polysaccharides can be thought of as a precipitation of system-power at the cellular level.

Polysaccharides could have a generalized signature, as in Equation 5.1.10, derived from the system-power family:

$$Sig_{polysaccharide} = Xa + \overline{Yb_{0-n}} \quad where \begin{bmatrix} X \in [S_{System_P}] \\ Y \in [S_{System_{P_T}}, S_{System_P}, S_{System_{\square}}, S_{System_N}] \\ a, b \text{ are integers}; a > b \end{bmatrix}$$

Eq 5.1.10: Polysaccharide

Glycogen is an example of a polysaccharide. Glycogen forms an energy reserve that can be quickly mobilized to meet a sudden need for glucose, but one that is less compact and more immediately available as an energy reserve than say triglycerides. Equation 5.1.11 for Glycogen follows:

$$Sig_{glycogen} = Xa + \overline{Yb_{0-n}} \quad where \begin{bmatrix} X \in [S_{System_P}] \\ Y \in [S_{System_{P_T}}, S_{System_P}, S_{System_K}, S_{System_N}] \\ a, b \; are \; integers; a > b \end{bmatrix}$$

Eq 5.1.11: Glycogen

Cellulose is another example of a polysaccharide. Cellulose is a polymer made with repeated glucose units bonded together by beta-linkages. Humans and many animals lack an enzyme to break the beta-linkages, so they do not digest cellulose. Equation 5.1.12 for Cellulose follows:

$$Sig_{cellulose} = Xa + \overline{Yb_{0-n}} \quad where \begin{bmatrix} X \in [S_{System_P}] \\ Y \in [S_{System_{P_T}}, S_{System_P}, S_{System_K}, S_{System_N}] \\ a, b \; are \; integers; a > b \end{bmatrix}$$

Eq 5.1.12: Cellulose

The primary element for the preceding polysaccharides X would be along the lines of 'energy storage'. The secondary elements may vary, with a Y element for Glycogen being 'rapid energy deployment' for example, and a Y element for Cellulose being 'bonded energy', for example.

Chapter 5.2: Sensations, Urges, Feelings, Thoughts

As human beings we experience sensations, urges and desires and wills, feelings and emotions, and thought. These are key aspects of our being and becoming and critical aspects of how choice at both the individual and collective levels may be determined.

This chapter will go over these fundamental aspects of self.

Sensations

Sensations are those things we experience with our senses. We see things, hear things, and smell things, taste things, can touch things. This ability to enter into relationship with objects through sensation is nothing other than a result of the emergence of Light's property of Presence. We become present to Presence through the device of sensation. Sensation can be thought of as the means by which this property of Light – Presence - molds or ingrains itself in us as human beings. Its potentiality, all which is contained in this aspect of Light, becomes available to us through the power of sensation. Hence an equation, Equation 5.2.1, will generally represent the family of sensations. Some elements that it would comprise of may be 'tangible', 'take notice of', amongst others.

$$Sig_{sensation} = Xa + \overline{Yb_{0-n}} \quad where \begin{bmatrix} X \in [S_{System_{Pr}}] \\ Y \in [S_{System_{Pr}}, S_{System_{P}}, S_{System_{K}}, S_{System_{N}}] \\ a, b \text{ are integers}; a > b \end{bmatrix}$$

Eq 5.2.1: Sensation

There could also be equations for hearing, seeing, tasting, touching, and smelling.

But there is also a deeper experience of sensation that is possible. When we see things, for instance, what are we seeing? Is it just the surface rendering of the play of matter, or do we see that the fullness of Light is still there, with all its potentiality and possibility, in the smallest thing we look at? Do we see that the whole universe and more is present in all its fullness in the least thing that we easily ignore, or belittle, or loathe? When we touch things is it the seeming concreteness of the play of the particles or atoms or chains of molecules that we touch? Or is it the Love and Light and the vastness of all that IS that allows itself to be as a small corner that we touch so as to make infinity be felt by something so finite?

Such a deeper contact offered through sensation suggests a subset of (5.2.1) with secondary elements perhaps described as 'fullness of Light', 'contacting infinity', amongst others, thus also yielding an equation form (5.2.2):

$$Sig_{deeper-sensation} = Xa + \overline{Yb_{0-n}} \quad where \begin{bmatrix} X \in [S_{System_{Pr}}] \\ Y \in [S_{System_{Pr}}, S_{System_P}, S_{System_K}, S_{System_N}] \\ a, b \text{ are integers}; a > b \end{bmatrix}$$

Eq 5.1.2: Deeper Sensation

Urges, Desires & Wills

Urges and desires and wills are similarly a play of the emergence of Light's property of Power. In the mystery of focus, the vastness of Light has projected itself in us into an apparent smallness that is in reality everything that is. And this smallness is trying through urge and desire and will to connect viscerally or even intentionally to other smallnesses that similarly are nothing other than the fullness of Light projected into a small smorgasbord of selected function. So the urge or desire for food, or companionship, or of possession, or of climbing a peak, is nothing other than Light's compressed property of Power, trying to reach more of the fullness that it is through a fulfillment of the urge or desire or will that it masquerades as. Hence urges can be represented as Equation 5.2.3:

$$Sig_{urges} = \frac{Xa + }{Yb_{0-n}} \quad where \begin{bmatrix} X \in [S_{System_P}] \\ Y \in [S_{System_{Pr}}, S_{System_P}, S_{System_K}, S_{System_N}] \\ a, b \text{ are integers}; a > b \end{bmatrix}$$

Eq 5.2.3: Urges

Elements may be of the type of 'grasp', 'possess', 'deeply connect', amongst others.

Feelings & Emotions

Feelings and emotions are a play of the emergence of Light's property of Harmony or Nurturing. Its instrument is the Heart and it generates an array of emotions that are an indication or active radar of whether we are moving toward or away from a reality of harmony, whether based on our small self or some larger Self of Light. Gradually, by navigating with these emotions and feelings we can get to a state where we always feel positive emotions which basically means we have more truly entered into relationship with some larger continent of Light. An equation for feelings is as represented by Equation 5.2.4:

$$Sig_{feelings} = Xa + \overline{Yb_{0-n}} \quad where \begin{bmatrix} X \in [S_{System_N}] \\ Y \in [S_{System_{Pr}}, S_{System_P}, S_{System_K}, S_{System_N}] \\ a, b \; are \; integers; a > b \end{bmatrix}$$

Eq 5.2.4: Feelings

Thoughts

Thoughts are a play of the emergence of Light's property of Knowledge. Through the thought we can become greater or conceptualize things greater or begun to enter into relationship with some things other than our small self. Thought allows us to connect to more "othernesses" or even the oneness of the reality of Light.

$$Sig_{thoughts} = Xa + \overline{Yb_{0-n}} \quad where \begin{bmatrix} X \in [S_{System_K}] \\ Y \in [S_{System_{Pr}}, S_{System_P}, S_{System_K}, S_{System_N}] \\ a, b \; are \; integers; a > b \end{bmatrix}$$

Eq 5.2.4: Thoughts

So we can see that the very substance of our becoming – sensations, urges, wills, desires, feelings, emotions, thoughts – are nothing other than an emergence of the four properties of light of presence, power, knowledge, and harmony.

Chapter 5.3: Four-Foldness at the Nation Level

A brief look at history will reinforce the idea that it is typically maturity along multiple and distinct dimensions, represented by the four properties of light, and their rich interaction that allows civilizations to endure.

Thus, those civilizations that have endured typically have a balance of all four families (Sri Aurobindo, 1971). Civilizations that have become extinct typically have had a focus on few drivers of innovation. Jared Diamond proposes five interconnected causes of collapse that may reinforce each other: non-sustainable exploitation of resources, climate changes, diminishing support from friendly societies, hostile neighbors, and inappropriate attitudes for change (Diamond, 2005). But these five sources may also be thought of as symptoms that arise due to the failure to adopt the catholicity of the sources of innovation emanating from each of the four sets or families. Further, the historian Toynbee suggested that societies decay because of their over-reliance on structures that helped them solve old problems (Toynbee, 1961). It can be interpreted that being thus biased they are unable to adopt the catholicity of the sources of innovation emanating from each of the four sets of families.

Approaching Civilization from a big-picture, global basis though, the sustainability of humankind will be ensured by a balance of development amongst the four sets of forces. This means that countries must be unique and in such a way that their primary emergence is distributed amongst all four sets of forces. Further, and based on this uniqueness, there must be an open and healthy interaction amongst these centers of uniqueness.

As an illustration, the uniqueness of representative countries in equation form may be as follows.

Example of Knowledge: India

Historically at least India, for example, in its essence, may be thought of as having an exceptional capacity for penetrating behind the surface, and further of meaningfully synthesizing many streams of development. Its primary power may thus be from the family of knowledge, with a strong secondary driver being its ability to create living harmonies. The equation for India will then likely be represented by Equation 5.3.1:

$$Sig_{India} = Xa + \overline{Yb_{0-n}} \ where \begin{bmatrix} X \in [S_{System_K}] \\ Y \in [S_{System_{Pr}}, S_{System_P}, S_{System_K}, S_{System_N}] \\ a, b \ are \ integers; a > b \end{bmatrix}$$

Eq 5.3.1: India (Knowledge Family)

Example of Power: Japan

Japan, in its essence, may be thought of as having a strong and noble warrior nature, along with a refined sense of aesthetics, amongst other qualities. Its equation, Equation 5.3.2 would be of the form:

$$Sig_{Japon} = Xa + \overline{Yb_{0-n}} \ where \begin{bmatrix} X \in [S_{System_P}] \\ Y \in [S_{System_{Pr}}, S_{System_P}, S_{System_K}, S_{System_N}] \\ a, b \ are \ integers; a > b \end{bmatrix}$$

Eq 5.3.2: Japan (Power Family)

Example of Nurturing: UK

UK, in its essence, may be thought of as having a strong ability to create practical, materialistic harmonies, resulting in such things as working parliaments and advanced democracy, for example. Its equation, Equation 5.3.3, would be of the form:

$$Sig_{UK} = Xa + \overline{Yb_{0-n}} \quad where \begin{bmatrix} X \in [S_{System_N}] \\ Y \in [S_{System_{P_T}}, S_{System_P}, S_{System_K}, S_{System_N}] \\ a, b \text{ are integers}; a > b \end{bmatrix}$$

Eq 5.3.3: UK (Nurturing Family)

Example of Service: Thailand

Thailand, in its essence, may be characterized by an exceptional sense of hospitality and sweet service, with an attention to detail in the practical arrangement of things. Its equation, Equation 5.3.4, would be of the form:

$$Sig_{Thailand} = Xa + \overline{Yb_{0-n}} \quad where \begin{bmatrix} X \in [S_{System_N}] \\ Y \in [S_{System_{P_T}}, S_{System_P}, S_{System_K}, S_{System_N}] \\ a, b \text{ are integers}; a > b \end{bmatrix}$$

Eq 5.3.4: Thailand (Service Family)

Emergence of Uniqueness

Similarly every nation on earth will have a uniqueness that can be represented by equations belonging to one of the four families.

As discussed in Chapter 2.5 on Emergence of Uniqueness, for uniqueness to emerge and mature is a process. The process is represented by Equation 2.5.1 reproduced here for convenience:

$$Sig_E = X \begin{bmatrix} C: Sig * mod\left(\int = 1\right) \\ F: Sig\ mod\ (c) \\ I: Sig\ mod\ \left(\int \overline{G, e, \pi}\right) \\ M: Sig * mod\ (G) \\ V: Sig * mod\ (e) \\ P: Sig * mod\ (\pi) \end{bmatrix}$$

Equations 5.3.1 – 4 represent example of the essence of uniqueness. For these to become living practicalities requires work at many different levels within a nation. If a nation has not really done the work to animate itself with its uniqueness then it may exist in the physical or P-state, or vital of V-state, or mental or M-state, which by definition lack sufficient maturity, and interaction with other nations is then going to be compromised. Moving to the integral or I-state, will allow a nation to at least not be locked in to a point of view. A consistent I-state practiced by each nation then is the minimum requirement for a sustainable global civilization.

Section 6: Complexification of Four-Foldness, Quantization, Super-Matter, & Expansion of the Universe

Matter is positioned as developing due to a process of four-fold complexification. This journey, summarizing the emergence of matter and life in previous chapters, appears in Chapter 6.1.

Chapter 6.2 and 6.3 describe the further complexification of matter by delving into the quantum realms. Quantum theory allows for the place of will or the observer in the determination of outcome. These chapters summarize such a will-focused process that may result in specific space, time, energy, and gravity quantization by which matter is changed.

Chapter 6.4 discusses the creation of super-matter. Super-matter is suggested to be that type of matter created through the intervention of conscious will or cohesive want. While complexification of matter as summarized in Chapter 6.1 is an "automatic" process coordinated by Nature, super-matter is a foundation based on will or cohesive want and sets the stage for a potentially unending development in which functional-richness existing in Light at Infinity can manifest in this material universe.

Chapter 6.5 draws a possible relationship between an expanding universe and the development of functionally-rich matter. As suggested creation of such matter requires space, time, energy, and gravity quantization which further implies that space as the prerequisite seeding ground of function must continue to expand.

Chapter 6.1: Complexification of Four-Foldness

As per discussion in Section 1, light slowing down to c resulted in the birthing of matter in the phenomenon of the Big Bang.

As per a common depiction of the history of the universe summarized by the Particle Data Group (Particle Data Group, 2015) at time, $t = 10^{-36}$ s, there was a high-intensity of light depicted by the cumulative energy of photons, $E = 10^{16}$ GeV. Existence of photons imply the existence of the EM Spectrum, and therefore tangibly, the existence of a four-fold wholeness projected from Light. As discussed previously in Section 3 this four-fold fullness was expressed by the wave-range of the EM Spectrum (Light's property of Knowledge), its implicit energy-gradient (Light's property of Power), the speed c with which is propagated (Light's property of Harmony), and the potential for mass (Light's property of Presence) also implicit in it.

In a 'Brief History of Time' Hawking (Hawking, 1988) states that the eventual goal of science is to provide a single theory that describes the whole universe. Scientists will usually separate this problem into two parts. First, focusing on laws that suggest how the universe changes with time. Second, focusing on the initial state of the universe. Hawking relates how some people feel that science should be concerned only with the first part. Since however, there appears to be very regular laws that govern how the universe changes with time, it must be equally reasonable to suppose that there are laws governing the initial state, he argues.

The existence of the four-foldness as an initial organizing principle is consistent with Hawking's argument. Only, in this mathematical model, it is the same set of laws that exists at both the initial starting point, and the subsequent development of the Universe, and as suggested in previous chapters.

In looking at the nature of the fourfold order it is apparent that the fourfold wholeness is more tightly related to a smaller set of fundamental physical attributes. Summarizing, as in Equations 6.1.1 through 6.1.4:

$$System_{Harmony} \propto c_U$$

Eq 6.1.1: System Harmony at Electromagnetic Level

$System_{Power} \propto hv$, where v is the frequency of the EM wave

Eq 6.1.2: System Power at Electromagnetic Level

$$System_{Knowledge} \propto [f(\lambda)]$$

Eq 6.1.3: System Knowledge at Electromagnetic Level

And since, $E = mc^2$ or $m = \frac{E}{c^2}$, and substituting hv for E,

$$System_{Presence} \propto {hv}/{c^2}$$

Eq 6.1.4: System Presence at Electromagnetic Level

But, $c \propto {1}/{h}$

Hence, the general relationships of the four-fold wholeness may be understood by knowing just two variables, c, and either, λ or v.

Now as per the model presented by the Particle Data Group at time, t = 10^{-10} s, the four-fold fullness begins to express itself in a series of particles. Quarks, leptons, bosons, which also may imply the presence of the Higgs-boson that gives quarks their mass are observed. But as discussed quarks are a precipitation of Light's property of Knowledge, leptons of Power, bosons of Harmony, and the Higgs-boson of Presence. Hence it is again observed that the four-fold fullness has expressed itself at a different order of complexity.

Einstein has suggested that particles are an excitation of the underlying field. The EM spectrum gives insight into a field that propagates at c at different frequencies. But there are other fields too. The Higgs Field has been suggested as being one of these (Jepsen, 2013). The Standard Model suggests that quarks, leptons, bosons, and the Higgs-boson are different fundamental particles, which implies that the fourfold wholeness has now a more complex basis of its operation, since as compared with the fourfold wholeness of the EM spectrum the number of independent basis appears to have increased as summarized in Figure 6.1.1, Complexification of Matter in Terms of Levels of Implicit Wholeness, to appear below.

Summarizing, as in Equations 6.1.5 through 6.1.8:

$$System_{Nurturing} \propto f(bosons)$$

Eq 6.1.5: System Nurturing at Particle Level

$$System_{Power} \propto f(leptons)$$

Eq 6.1.6: System Power at Particle Level

$$System_{Knowledge} \propto f(quarks)$$

Eq 6.1.7: System Knowledge at Particle Level

$$System_{Presence} \propto f(Higgs\_boson)$$

Eq 6.1.8: System Presence at Particle Level

As discussed in Chapter 4.3 the Periodic Table itself is also an expression of the four-fold wholeness. Hence, d-group elements are an instance of Light's property of Presence, p-group elements are an instance of Light's property of Knowledge, s-group elements are an instance of Light's property of Power, and f-group elements are an instance of Light's property of Harmony or Nurturing. As per the Particle Data Group model (Particle Data Group, 2015) this emergence likely began at $t = 3 \times 10^5$ years when the atom is suggested to have emerged, and continued at least to $t = 10^9$ years with the emergence of stars which it is already known are the furnaces in which heavier atoms were created.

Leveraging off the fundamentally different probability clouds, as in Equations 6.1.9 through 6.1.12:

$$System_{Nurturing} \propto f(f\_orbital)$$

Eq 6.1.9: System Nurturing at Atomic Level

$$System_{Power} \propto f(s\_orbital)$$

Eq 6.1.10: System Power at Atomic Level

$$System_{Knowledge} \propto f(p\_orbital)$$

Eq 6.1.11: System Knowledge at Atomic Level

$$System_{Presence} \propto f(d\_orbital)$$

Eq 6.1.12: System Presence at Atomic Level

Atoms therefore have at least three four-fold wholenesses implicit in them.

At time, $t = 13.8 \times 10^9$ years, there is a fourth clear expression of the same four-fold order as the bases of an even more complex organization, that of cellular life and all that is founded on it. This can be summarized as in Equations 6.1.13 through 6.1.16:

$$System_{Nurturing} \propto f(lipids)$$

Eq 6.1.13: System Nurturing at Cellular Level

$$System_{Power} \propto f(polysaccharides)$$

Eq 6.1.14: System Nurturing at Cellular Level

$$System_{Knowledge} \propto f(nucleic\ acids)$$

Eq 6.1.15: System Knowledge at Cellular Level

$$System_{Presence} \propto f(proteins)$$

Eq 6.1.16: System Presence at Cellular Level

Cellular life therefore seems to have at least four four-fold-wholenesses implicit in it to therefore further complexify the architecture of matter.

This scheme of complexification is summarized by Figure 6.1.1:

Level	Independent "Variables"	Minimum Levels of Implicit Wholeness	Minimum Number of Levels
EM Spectrum	C, ν *or* λ	• EM	1
Particle Level	Quarks, Leptons, Bosons, Higgs-Boson	• EM • Particle	2
Atomic Level	S-orbital, P-orbital, D-orbital, F-orbital	• EM • Particle • Atom	3
Cellular Level	Proteins, Nucleic Acids, Lipids, Polysaccharides	• EM • Particle • Atom • Cell	4

Figure 6.1.1 Complexification of Matter in Terms of Levels of Implicit Wholeness

In reference to the four sets at M_2 (as discussed in Chapter 2.3 on Architectural Forces) it may also be that as the basis of matter complexifies as it journeys through the EM spectrum and other potential fields, the elementary particles, the atom, and cellular life, as summarized in Figure 6.1.1, the sets which have been positioned to each contain infinite elements, concretely manifest more of their function-elements.

This suggestion has been the basis of the equations derived in previous chapters. The complexity of matter may therefore be related to the number of manifested function-elements of the set of four sets. This idea is consistent with the notion of the "adjacent possible" suggested by Kaufmann (Kaufmann, 2003) in which innovation is positioned as a recombination of existing parts to create new value – or of existing sets to combine parts of themselves to create new elements based on new circumstance. If MS signifies manifested-set, so that the cardinality or number of elements in the combined set is the union of the four manifested-sets, this may be summarized by the following equation, Equation 6.1.17:

$$Matter_{Complexity} \propto \left| MS_{System_{Pr}} \cup MS_{System_{P}} \cup MS_{System_{K}} \cup MS_{System_{N}} \right|$$

Eq 6.1.17: Complexity of Matter in Terms of Manifested Set

Chapter 6.2: Quantization and Quantum-Certainty at the Cellular Level

The quantum is a window into layers of reality behind the surface layer U. This insight was captured by Schrodinger's Equation (Chapter 2.10) and by Heisenberg's Uncertainty Principle (Chapter 2.11), the former suggesting that matter, or what is going to appear materially, is a function of some incredible number of superposed states containing infinite possibility, and the latter suggesting that quantum fluctuation due to an always buzzing pregnant-infinity, lies behind everything even when seemingly still.

The Light-Space-Time Emergence equation (2.8.3) derived in Chapter 2.8, and reproduced here for convenience, may provide some insight into the dynamics of the pregnant-infinity, and of quantum-certainty in the adaptabilities of the cell. Recall that the basis of cell adaptability was discussed in Chapter 5.1. Reproducing the equation:

$$
Emergence_{light-space-time} = \left\| \begin{array}{cc}
\left[\begin{array}{c}
C_{\infty}: [Pr, Po, K, H] \\
(\downarrow R_{C_K} = R_{C_{\infty}}) \\
C_K: [S_{Pr}, S_{Po}, S_K, S_H] \\
(\downarrow R_{C_N} = R_{C_K}) \\
C_{N:} f(S_{Pr}\ x\ S_{Po}\ x\ S_K\ x\ S_H) \\
(\downarrow R_{C_U} = R_{C_N)} \\
C_U: [P, V, M, C]
\end{array} \right]_{Light}
&
\left[\begin{array}{c}
M_3 \rightarrow System_X \\
(\uparrow F \rightarrow I) \\
M_2 \rightarrow S_{System_X} \\
(\uparrow Sig \rightarrow F) \\
M_1 \rightarrow Sig_x \\
(\uparrow > P_{x)} \\
U \rightarrow x_U
\end{array} \right]_{Space}
\\
\left[\begin{array}{c}
M_3 : All\ t \\
\downarrow \\
M_2 : 0 \geq t > \infty \\
\downarrow \\
M_1 : 0 > t > \infty \\
\downarrow \\
U \rightarrow \begin{array}{c} t \sim E_{Cell};\ \mathrm{TC}: M_3 \rightarrow \mathrm{U} \\ t \sim E_{Human};\ \mathrm{TC}: U \rightarrow M_3 \end{array}
\end{array} \right]_{Time}
&
TC \rightarrow x_T
\end{array} \right\|_{\langle x_U | x_T \rangle}
$$

The Light-Matrix, the top-left matrix in (2.8.3), suggests the possibilities in the pregnant-infinity. These possibilities, recall, have been set up by realities so created by light traveling at different speeds. While prevalent dynamics at the surface layer, U, are always influenced by dynamics from the deeper layers, there is a state that can be created that will allow a more focused and intentional activation of the meta-functions resident in the deeper layers. Such activation will potentially trigger a series of processes culminating in a state of quantum-certainty as will be explored through the rest of this chapter.

The Space-Matrix, the top-right matrix in (2.8.3), suggests the dynamics involved in creating such an activation-state. Essentially this involves the overcoming of patterns at U. Consider the case of a hypothetical insect, for example, that is always prey to other predators. At some point a visceral urge to overcome some of these predators may arise in the insect. This visceral urge is a breaking of habitual patterns common to that species of insect. If it is deep enough and pervasive enough, it can be thought of as an activation-state and will allow the creation of a quantum-window so that at every micro level there is now a connection with the quantum worlds, Q, behind. It is proposed that for effective adaptability to come about it is only such a pervasive state that will open a quantum-window to effectively stimulate the interaction between layers that will ultimately allow adaptability to occur. This activation-state can be further specified by leveraging (2.7.4) – The Generalized Equation for Organizational Direction, derived in Chapter 2.7, on Qualified Determinism.

Reproducing Equation 2.7.4:

$$Org\_Dir = DI\left(\begin{bmatrix} M_3 \rightarrow System_X \\ (\uparrow F \rightarrow I) \\ M_2 \rightarrow S_{System_X} \\ (\uparrow Sig \rightarrow F) \\ M_1 \rightarrow Sig_x \\ (\uparrow > P_{P)} \\ U \rightarrow x_U \end{bmatrix}_{x=p,v,m,i}\right) \rightarrow$$
$$x\_matrix_{strongest} @ level_{strongest}$$

The activation-state can be thought of as being invoked when $level_{strongest}$ is at least M_1. Hence, as in Equation 6.2.1:

$$Activation - State_{condition} \geq M_1$$

Eq 6.2.1: Condition for Activation-State

Note that the Time-Matrix, in the lower-left side of (2.8.3), specifies that approximately, adaptability may tend to be more automatic up to the emergence of cellular organisms. This is specified by the direction

$M_3 \rightarrow$ U, which indicates that it is the meta-level that organizes activity at U. At the human level the direction is flipped as specified by $U \rightarrow M_3$, and suggests that will, intention, feeling and the like are more important in stimulating the process of adaptability.

The visceral urge, then, may stimulate interaction with a collective-intelligence or 'specific-species-intelligence meta-function' that allows the generation of a new and specific 'predator-overcoming meta-function' so that the hypothetical insect in question goes through an adaptation to survive at least some kinds of predator attacks. This may take the form of this species of insect creating a hard-shell around it. Mathematically this new meta-function may take the following form as suggested by Equation 6.2.2:

$$Sig_{hard-protective-shell} = Xa + \overline{Yb_{0-n}} \ \ where \left[\begin{array}{c} X \in [S_{System_P}] \\ Y \in [S_{System_{Pr}}, S_{System_P}, S_{System_K}, S_{System_N}] \\ a, b \ are \ integers; a > b \end{array} \right]$$

Eq 6.2.2: Hard Protective Shell Meta-Function

Specifically, S_{System_P} relates to the set of Polysaccharides and suggests that the chains of sugar molecules will adapt to become a shell to protect the insect. The primary element X, is therefore an element of the set or Polysaccharides. $S_{System_{Pr}}$ refers to the set of Proteins. S_{System_K} refers to the set of Nucleic Acids. S_{System_N} refers to the set of Lipids. Y as the secondary element will invoke the action of some proteins, some existing polysaccharides, some nucleic acids, and some lipids in bringing about the adaptation as specified by $Sig_{hard-protective-shell}$. Perhaps the nucleic acids will code or coordinate how the proteins will work with the existing polysaccharides and lipids to create a new arrangement of polysaccharides that becomes the protective layer for the insect. This new arrangement of polysaccharides then becomes the purpose of a particular new type of specialized cell that exists to protect the insect against certain types of predators.

So while the breaking of patterns as specified by the Space-Matrix was a first step in the process, the subsequent creation of a new meta-function can be thought of as a second step. This new meta-function can be thought of as happening due to some combined functioning of the Light and Space Matrices. Specifically in the Space-Matrix, the breaking of patterns, $(\uparrow > P_x)$, allows the dynamics of M_1 to become active: $M_1 \rightarrow Sig_x$. This allows the ever-present layers of light, and specifically $C_{N:} f(S_{Pr}\ x\ S_{Po}\ x\ S_K\ x\ S_H)$ in the Light-Matrix to become active so that new meta-function is generated.

The third step is suggested by $(\downarrow R_{C_U}{}' = R_{C_N})$ that allows specific quantization to occur. It is proposed that the hard-protective-shell adaptability becomes real through a series of tightly coordinated space, time, energy, and gravity quantization as specified by (2.12.2-5) derived in Chapter 2.12. These equations are reproduced here for convenience and suggest the mechanics for the adaptability along this specific line of development in the larger scheme of Natural History.

Equation 2.12.2 models space-quantization:

$$Space_{quantization} = h_{UK}(Xa + \overline{Yb_{0-n}})$$

$$where: \begin{bmatrix} X \in [S_{System_K}] \\ Y \in [S_{System_{Pr}}, S_{System_P}, S_{System_K}, S_{System_N}] \\ a, b\ are\ integers; a > b \end{bmatrix}$$

Recall that space-quantization further structures space so that seeds of knowledge or knowledge-potential to do with the new meta-function are now resident in the material ecosystem associated with the meta-function. In such a manner the species that set this meta-function into action will more and more tap into the possibilities to bring about the new protective mechanism it seeks.

Equation 2.12.3 models time-quantization:

$$Time_{quantization} = h_{UP}(Xa + \overline{Yb_{0-n}})$$

$$where: \left[\begin{array}{c} X \in [S_{System_P}] \\ Y \in [S_{System_{Pr}}, S_{System_P}, S_{System_K}, S_{System_N}] \\ a, b \text{ are integers}; a > b \end{array}\right]$$

Recall that time-quantization alters the structure of time so that phases of maturity associated with seeds in space are now encoded in time. Slow and faster periods of change are in this way connected to the species seeking to protect itself against predators.

Equation 2.12.4 models energy-quantization:

$$Energy_{quantization} = h_{UPr}(Xa + \overline{Yb_{0-n}})$$

$$where: \left[\begin{array}{c} X \in [S_{System_{Pr}}] \\ Y \in [S_{System_{Pr}}, S_{System_P}, S_{System_K}, S_{System_N}] \\ a, b \text{ are integers}; a > b \end{array}\right]$$

Recall that energy-quantization allows matter to be formed, and in this case, will have to do with the incremental material changes that the species will go through in forming a hard protective shell.

Equation 2.12.5 models gravity-quantization:

$$Gravity_{quantization} = h_{UN}(Xa + \overline{Yb_{0-n}})$$

$$where: \begin{bmatrix} X \in [S_{System_N}] \\ Y \in [S_{System_{Pr}}, S_{System_P}, S_{System_K}, S_{System_N}] \\ a, b \text{ are integers}; a > b \end{bmatrix}$$

Recall that gravity-quantization has to do with inter-relation between the species and surrounding objects and will change the very nature of gravity to allow a subtle new balance in the species interaction with its surrounding so that the deep urge of the species can more easily be fulfilled.

Such quantization of space, time, energy, and gravity, describes the state of quantum-certainty. Quantum-certainty is the result of an intimate interaction with Q, the worlds accessed through quanta. The process of quantization and the state of quantum-certainty initiated through an activation-state, and the consequent enhancement or creation of a new meta-function, give a more intimate sense of the tightly coherent system we live in. Urge or desire, if deep enough, can set into motion an inter-relation between easily distinct layers of existence, allowing a deeper 'system-being' to intervene in regular system functioning.

Chapter 6.3: Quantization and Quantum-Certainty at the Human-Level

Just as matter can be further complexified to accommodate enhanced functionality as described in the example of an insect in the previous chapter, so too will matter continue to change with the application of human will and want. This chapter describes the process of quantization by which this might occur.

The Light-Space-Time Emergence equation (2.8.3) derived in Chapter 2.8, and reproduced here for convenience, will provide some insight into the dynamics of the pregnant-infinity, and of quantum-certainty in the making of history. Reproducing the equation:

$$Emergence_{light-space-time} = \left\| \begin{matrix} \begin{bmatrix} C_\infty: [Pr, Po, K, H] \\ (\downarrow R_{C_K} = R_{C_\infty}) \\ C_K: [S_{Pr}, S_{Po}, S_K, S_H] \\ (\downarrow R_{C_N} = R_{C_K}) \\ C_N: f(S_{Pr}\ x\ S_{Po}\ x\ S_K\ x\ S_H) \\ (\downarrow R_{C_U} = R_{C_N}) \\ C_U: [P, V, M, C] \end{bmatrix}_{Light} & \begin{bmatrix} M_3 \rightarrow System_X \\ (\uparrow F \rightarrow I) \\ M_2 \rightarrow S_{System_X} \\ (\uparrow Sig \rightarrow F) \\ M_1 \rightarrow Sig_x \\ (\uparrow > P_x) \\ U \rightarrow x_U \end{bmatrix}_{Space} \\ \begin{bmatrix} & M_3 : All\ t \\ & \downarrow \\ & M_2 : 0 \geq t > \infty \\ & \downarrow \\ & M_1 : 0 > t > \infty \\ & \downarrow \\ U \rightarrow & t \sim E_{Cell};\ TC: M_3 \rightarrow U \\ & t \sim E_{Human};\ TC: U \rightarrow M_3 \end{bmatrix}_{Time} & TC \rightarrow x_T \end{matrix} \right\|_{\langle x_U | x_T \rangle}$$

The Light-Matrix, the top-left matrix in (2.8.3), suggests the possibilities in the pregnant-infinity. These possibilities, recall, have been set up by realities so created by light traveling at different speeds. While prevalent dynamics at the surface layer, U, are always influenced by dynamics from the deeper layers, there is an activation-state that can be created that will allow a more focused and intentional activation or creation of the meta-functions in the deeper layers.

The Space-Matrix, the top-right matrix in (2.8.3), suggests the dynamics involved in creating such an activation-state. Essentially this involves the overcoming of patterns at U. Note that by definition such an activation-state necessitates moving in the direction where the speed of light increases. Note that speed of light increasing implies that the sense of separation is diminishing, and entities can have a more complete access to the fullness that they are. This can only happen when patterns get elevated. If patterns are in the U realm, being therefore untransformed, that will by definition not allow a quantum-window to open.

As per discussion in Chapter 6.2, this activation-state can be further specified by leveraging (2.7.4) – The Generalized Equation for Organizational Direction, derived in Chapter 2.7, on Qualified Determinism.

Reproducing Equation 2.7.4:

$$Org\_Dir = DI\left(\begin{bmatrix} M_3 \rightarrow System_X \\ (\uparrow F \rightarrow I) \\ M_2 \rightarrow S_{System_X} \\ (\uparrow Sig \rightarrow F) \\ M_1 \rightarrow Sig_x \\ (\uparrow > P_P) \\ U \rightarrow x_U \end{bmatrix}_{x=p,v,m,i}\right) \rightarrow$$

$$x\_matrix_{strongest} \; @ \; level_{strongest}$$

Activation-state can be thought of as being invoked when $level_{strongest}$ is at least M_1. Hence, as suggested by the previously derived (6.2.1), reproduced here for convenience:

$Activation - state_{condition} \geq M_1$

Further, and as per the discussion in the previous chapter, assume that there is a deep wanting felt by a threshold number of people. Perhaps this wanting, a step towards a global sustainable civilization, is for each nation to operate more from their true essence, or I-state, as opposed to fundamentalist form. In other words a meta-function, $Sig_{Nation-essence}$, must be created that will then organize its own reality, and as indicated by Equation 6.3.1.

$$Sig_{Nation-essence} = Xa + \overline{Yb_{0-n}}$$

$$where \begin{bmatrix} X \in [S_{System_K}] \\ Y \in [S_{Systemp_{Pr}}, S_{Systemp_P}, S_{System_K}, S_{System_N}] \\ a, b \text{ are integers}; a > b \end{bmatrix}$$

Eq 6.3.1: Nation-Essence Meta-Function

While the key aspect, or primary X-element in the nation-essence meta-function may be an element from the set of Knowledge, and specifically 'what the nation-essence means', there will be many parts, represented by the Y-elements, such as the structure, processes, institutions, culture, that will also need to be created. An important denominator in breaking patterns in each of these areas though is hinted at in Chapter 4.2, and involves the creation of 'cohesion' at the individual and collective levels as in Equation 6.3.2, Nation-Essence-Cohesion with the primary element belonging to the set of Harmony or Nurturing:

$$Sig_{Nation-essence-cohesion} = Xa + \overline{Yb_{0-n}}$$

$$where \begin{bmatrix} X \in [S_{System_N}] \\ Y \in [S_{System_{Pr}}, S_{System_P}, S_{System_K}, S_{System_N}] \\ a, b \text{ are integers}; a > b \end{bmatrix}$$

Eq 6.3.2: Nation-Essence-Cohesion Meta-Function

Hence, S_{System_K} relates to the set of Thoughts required to contextualize and frame nation-essence. $S_{System_{Pr}}$ refers to the set of Sensations required to assess and interpret the constantly arising signs of the new development – the sensory cues as it were that will allow any individual or collectivity to sense that they were on the right path. S_{System_P} refers to the set of Urges and Wills constantly required to ensure that the goal of cohesion was attained. S_{System_N} refers to the set of Feelings and Emotions that will need to be generated to attain cohesion.

As such cohesion becomes a reality, older patterns will break down and there will be easier and longer activation-state periods, potentially allowing 'miracle' to become the law.

As suggested in Chapter 6.2, while the breaking of patterns as specified by the Space-Matrix was a first step in the process, the subsequent creation of a new meta-function can be thought of as a second step. This new meta-function can be thought of as happening due to some combined functioning of the Light and Space Matrices. Specifically in the Space-Matrix, the breaking of patterns, $(\uparrow > P_x)$, allows the dynamics of M_1 to become active: $M_1 \rightarrow Sig_x$. This allows the ever-present layers of light, and specifically $C_{N:} f(S_{Pr}\ x\ S_{Po}\ x\ S_K\ x\ S_H)$ in the Light-Matrix to become active so that new meta-function is generated.

The third step is suggested by ($\downarrow R_{C_U} = R_{C_N}$) that allows specific quantization to occur. Quantization is of fundamental importance because it is what allows the fabric of experienced reality to change. The new thought to do with nation-essence is quantized as seeds of knowledge in space, fundamentally changing the structure of space. Hence, reproducing Equation 2.12.2, space-quantization is modeled as:

$$Space_{quantization} = h_{UK}\left(Xa + \overline{Yb_{0-n}}\ \right)$$

$$where: \begin{bmatrix} X \in [S_{System_K}] \\ Y \in [S_{System_{Pr}}, S_{System_P}, S_{System_K}, S_{System_N}] \\ a, b\ are\ integers; a > b \end{bmatrix}$$

The phases of maturity that the seeds of knowledge will go through give time a new reality or structure. In other words time too is re-oriented to promote the outcome of the new meta-function in effect. This quantization of time, essentially giving time in this new fabric a different meaning is modeled by Equation 2.12.3, reproduced here for convenience:

$$Time_{quantization} = h_{UP}\left(Xa + \overline{Yb_{0-n}}\ \right)$$

$$where: \begin{bmatrix} X \in [S_{System_P}] \\ Y \in [S_{System_{Pr}}, S_{System_P}, S_{System_K}, S_{System_N}] \\ a, b\ are\ integers; a > b \end{bmatrix}$$

The new meta-function changes the very nature of matter, which is quantized to allow the material fabric of existence to promote the new meta-function in unimaginable ways. This quantization of energy, resulting in a subtly different matter is modeled by Equation 2.12.4, reproduced here for convenience:

$$Energy_{quantization} = h_{UPr}\left(Xa + \overline{Yb_{0-n}}\ \right)$$

$$where: \begin{bmatrix} X \in [S_{System_{Pr}}] \\ Y \in [S_{System_{Pr}}, S_{System_{P}}, S_{System_{K}}, S_{System_{N}}] \\ a, b \text{ are integers}; a > b \end{bmatrix}$$

As a result the very force of gravity is altered locally as well, so that its quantization promotes subtle interaction in the related ecosystem that will be predisposed to making the meta-function a reality. This quantization of gravity is modeled by Equation 2.12.5, reproduced here for convenience:

$$Gravity_{quantization} = h_{UN}(Xa + \overline{Yb_{0-n}})$$

$$where: \begin{bmatrix} X \in [S_{System_{N}}] \\ Y \in [S_{System_{Pr}}, S_{System_{P}}, S_{System_{K}}, S_{System_{N}}] \\ a, b \text{ are integers}; a > b \end{bmatrix}$$

It is the combined quantization that can be thought of as quantum-certainty. Such a process of quantum-certainty has the effect of changing the vary fabric or make-up of matter.

Chapter 6.4: Transitioning to Super-Matter

Super-Matter is positioned as being fundamentally different from matter. While matter can be thought of as the result of Nature's automatic working, super-matter can be thought of as the result of a conscious will and cohesive wanting that causes a process of quantization by which the very fabric of matter is changed. As the previous chapters suggest, such quantization is related to enhancing the functional-richness of matter. Imagine through a cohesive-wanting matter being able to become more flexible, more durable, more stretchable, light at will, heavy at will, change color at will, expand in size, contract in size, amongst infinite other functions and possibilities contained in Light at infinity.

The previous chapters, 6.2 and 6.3, looked at the micro-level process by which quantization changes the multi-dimensional space-time-energy-gravity fabric of material existence. This chapter takes a more macro-level view to summarize three distinct phases of matter: that of ordinary-matter, matter-in-transition, and super-matter. In the process the requirement for an age of super-matter is made clearer.

The Light-Space-Time Emergence equation (2.8.3) derived in Chapter 2.8, and reproduced here for convenience, will be leveraged to derive summary equations to distinguish between the different phases of matter. Reproducing the equation:

$$
Emergence_{light-space-time}
= \left| \begin{array}{cc}
\begin{bmatrix}
C_{\infty}: [Pr, Po, K, H] \\
(\downarrow R_{C_K} = R_{C_{\infty}}) \\
C_K: [S_{Pr}, S_{Po}, S_K, S_H] \\
(\downarrow R_{C_N} = R_{C_K}) \\
C_{N:}\ f(S_{Pr}\ x\ S_{Po}\ x\ S_K\ x\ S_H) \\
(\downarrow R_{C_U} = R_{C_{N)}} \\
C_U:\ [P, V, M, C]
\end{bmatrix}_{Light}
&
\begin{bmatrix}
M_3 \rightarrow System_X \\
(\uparrow F \rightarrow I) \\
M_2 \rightarrow S_{System_X} \\
(\uparrow Sig \rightarrow F) \\
M_1 \rightarrow Sig_x \\
(\uparrow > \dot{P}_{x)} \\
U \rightarrow x_U
\end{bmatrix}_{Space}
\\
\begin{bmatrix}
 & M_3 : All\ t \\
 & \downarrow \\
 & M_2 : 0 \geq \mathrm{t} > \infty \\
 & \downarrow \\
 & M_1 : 0 > \mathrm{t} > \infty \\
 & \downarrow \\
U \rightarrow & \mathrm{t} \sim E_{Cell};\ \mathrm{TC}: M_3 \rightarrow \mathrm{U} \\
 & \mathrm{t} \sim E_{Human};\ \mathrm{TC}: U \rightarrow M_3
\end{bmatrix}_{Time}
& TC \rightarrow x_T
\end{array} \right|_{\langle x_U | x_T \rangle}
$$

Simplifying each of the main matrices yields Equation 6.4.1, the simplified version of the Light-Space-Time Emergence equation:

$$Emergence_{light-space-time} = |[L][S][T]TC \rightarrow x_T|_{\langle x_U | x_T \rangle}$$

Eq. 6.4.1: Simplified Version of Light-Space-Time Emergence

As explained by Neil Turok in his book The Universe Within (Turok, 2012), Euler's formula reproduced below, can be used to model many naturally occurring phenomena because of its sinusoidal oscillation between narrow bounds as x increases. Reproducing Euler's formula:

$$e^{ix} = \cos x + i \sin x$$

The sum of the squares of the ordinary and complex parts, on the right side of the equation, is one. In quantum theory this ensures that the probabilities for all possible outcomes add up to one. Hence this formula is useful when considering the macro-level effects of quantization of the fabric of material existence.

Further, in modeling matter a modified notation of the Schrodinger wavefunction as interpreted by Feynman is leveraged. This version features the integral sign, $\int$, meaning that all terms to the right of it have to be summed up for all space and time till the moment when the wavefunction is required to be known.

Hence, combining (6.4.1) with Feynman's interpretation of the Schrodinger wavefunction, with the Euler formula yields the following equations, 6.4.2-7 for matter:

$$\psi_{established-matter} = \left| \int e^{i \int |[L][S][T]|} \right|_U$$

Eq. 6.4.2: Wavefunction for Established-Matter

(6.4.2) suggests that so long as the basis of matter is untransformed, specified by the U following the vertical-brackets, the outcome is not going to be any different from established matter, specified by $\psi_{established-matter}$.

Further, as specified by Equation 6.4.3, the Probability-View of Established-Matter, the very basis of matter is going to be either the untransformed physical (P_U), the untransformed vital (V_U), the untransformed mental (M_U), or the untransformed integral (I_U):

$$|\psi_{established-matter}|^2 = P_U^2 + V_U^2 + M_U^2 + I_U^2 = 1$$

Eq. 6.4.3: Probability-View for Established-Matter

As specified by (6.4.3) the probability that any of these bases will be leveraged in the creation of matter adds up to one. Note that the physical, the vital, the mental, and integral were discussed in Chapter 2.6 on Inherent Dynamics of a System.

Equation 6.4.4, Wavefunction for Matter-in-Transition, suggests the mixed bases for matter, as specified by dynamics of both the untransformed (U) and the meta-levels (M_x), which therefore results in matter-in-transition.

$$\psi_{matter-in-transition} = \left| \int e^{i \int |[L][S][T]|} \right|_{U\&M_x}$$

Eq. 6.4.4: Wavefunction for Matter-in-Transition

As specified by Equation 6.4.5, Probability-View for Matter-in-Transition, the probability that any of the untransformed (U) and transformed (T) bases will be leveraged in the creation of matter adds up to one.

$$|\psi_{matter-in-transition}|^2 = P_U^2 + P_T^2 + V_U^2 + V_T^2 + M_U^2 + M_T^2 + I_U^2 + I_T^2 = 1$$

Eq. 6.4.5: Probability View for Matter-in-Transition

Equation 6.4.6, Wavefunction for Super-Matter, suggests some transformed bases for matter, as specified by dynamics of the meta-levels (M_x), which therefore results in super-matter.

$$\psi_{super-matter} = \left| \int e^{i \int |[L][S][T]|} \right|_{M_x}$$

Eq. 6.4.6: Wavefunction for Super-Matter

As specified by Equation 6.4.7, Probability-View for Super-Matter, the probability that the transformed (T) bases will be leveraged in the creation of matter adds up to one.

$$\left|\psi_{super-matter}\right|^2 = P_T^2 + V_T^2 + M_T^2 + I_T^2 = 1$$

Eq. 6.4.7: Probability View for Super-Matter

Chapter 6.5: Expansion of the Universe

In previous chapters it is suggested that a four-fold quantization is required for the fabric of material existence to change. Further in Chapters 6.2 and 6.3 it is suggested that a conscious and cohesive want for added functional-richness at the material level can trigger a process culminating in quantum-certainty in which such quantization occurs. In Chapter 6.4 a differentiation is made between three types of matter – ordinary-matter, matter-in-transition, and super-matter. Development of functional richness precipitating in a change of matter can happen in either of these types of phases. In ordinary-matter it is automatic though through the coordination of Nature. In matter-in-transition there may be cohesive want for added functional richness that requires the overcoming of habitual patterns. But equally, material reality can be prolonged through habitual patterns hence broadening the bases for functional-enhancement to include both untransformed function and transformed function. In super-matter a cohesive wanting implies the consistent overcoming of habitual patterns, and the cumulative effect at the macro-level is that a foundation of super-matter is laid which then sets in place a virtuous cycle as suggested by the associated Schrodinger wavefunction for super-matter as explored in Chapter 6.4.

If the cumulative effect of overcoming habitual patterns increases the likelihood of super-matter being formed, this implies the increase too of four-fold quantization. Space, as the seeding ground of knowledge, will therefore also require expansion to continue to allow seeding to take place. But if space needs to expand in such a scenario, then the fact that scientists have observed that the universe is expanding may be related to this. This chapter explores a possible link between the expansion of the universe and an increase in functional-richness that implies the expansion of space.

The entertainment of such an idea perhaps can be strengthened through consideration of the cosmological principle that suggests that the spatial distribution of matter in the universe is homogenous and isotropic when viewed on a threshold scale of 250 million light years. The nine-year cosmic microwave background image created with the Wilkinson Microwave Anisotropy Probe (WMAP) depicts 13.77 Billion year old temperature differences that correspond to the seeds that became the galaxies (NASA-WMAP, 2014). This homogeneity in temperature difference in the early universe suggests too a process of homogenous four-fold quantization that set into motion the development of galaxies.

Complexity of Matter, Dark Matter, & Dark Energy

According to NASA only 5% of the matter in the universe is visible, while as much as 27% is 'dark matter'. It is the 5% of 'visible' matter though that forms the familiar stars, galaxies, or the atoms (NASA-darkmatter, 2016). But since, as explored in Chapter 2.7 on Qualified Determinism, the architectural sets at M_2 are infinite the question is why would all functions need to manifest in the same way?

As stated in an article 'Dark Energy: The Biggest Mystery in the Universe' (Panek, 2010), "Sight itself has blinded us to the Universe". It should be possible for other matter-based constructions to come into being based on the functionality they exist for. For instance, as already suggested there could be other fields that create other kinds of elementary particles, that create other kinds of atoms and that results in other kinds of structures not visible with our current instrumentation.

Recall (6.1.17) on complexity of matter in terms of the manifested sets:

$$Matter_{Complexity} \propto \left| MS_{System_{Pr}} \cup MS_{System_{P}} \cup MS_{System_{K}} \cup MS_{System_{N}} \right|$$

It is also possible to state an Equation, 6.5.1, relating complexity of matter in terms of visible and dark matter:

$$Matter_{Complexity} \propto [Visible\ Matter + Dark\ Matter]$$

Eq 6.5.1: Complexity of Matter in Terms of Visible and Dark Matter

Further, as suggested by the previous discussions on the quantization of space in Section 1 and the earlier chapters of Section 6 space is not empty but at the very least is a seeding ground for a variety of emergences. The notion of space having 'amazing' properties was, according to a report on 'dark energy' by the Harvard-Smithsonian Center for Astrophysics, first suggested by Einstein (Harvard-Smithsonian Center for Astrophysics, 2004). Quoting: "Einstein was the first person to realize that empty space is not nothingness. Space has amazing properties, many of which are just beginning to be understood." Einstein had suggested the existence of a 'dark energy' about 100 years ago (NASA-Supernova, 2001) as a property of space that caused the expansion of the universe. Dark Energy is estimated to comprise as much as 68% of the universe.

Reinterpretation of Cosmological Expansion-Contraction Dynamics

Keeping in mind the complexification of matter as it journeys from the field-level through the quantum-, atomic-, and cellular-levels, summarized in Figure 6.5.1, it may be possible to re-interpret the supposed expansion-contraction dynamics of cosmology in relation to the mathematical model presented in this book.

Principle	Field Level	Quantum Level	Atomic Level	Cellular Level
System-knowledge	EM Spectrum	Quarks	p-group	Nucleic Acids
System-presence	Mass	Higgs-boson	d-group	Proteins
System-power	Energy-gradient implicit in EM Spectrum	Leptons	s-group	Polysaccharides
System-nurturing	C, speed of EM Spectrum or Light	Bosons	f-group	Lipids

Figure 6.5.1 Complexification of Matter Related to Four-Fold Principle

First, flipping the left and right sides of the equation on $Matter_{Complexity}$ to yield Equation 6.5.2:

$$\left|MS_{System_{Pr}} \cup MS_{System_P} \cup MS_{System_K} \cup MS_{System_N}\right| \propto Matter_{Complexity}$$

Eq 6.5.2: Flipped Matter-Complexity Equation

This implies that the manifested-set, MS, is growing at a certain threshold level. Assuming this threshold level, $MS\_Growth_{Threshold}$, is a property of space related to dark energy it may be possible to restate the condition of cosmological expansion and contraction. Hence, so long as the MS is increasing at a certain rate that exceeds $MS\_Growth_{Threshold}$ the level of dark energy is such that the acceleration of galaxies, $Acceleration_{Galaxies}$ exceeds the contracting force of gravity at the universal level, $Gravity_{Universe}$.

Therefore, as in Equations 6.5.3 and 6.5.4:

$$MS\_Growth_{Threshold}\text{: } Acceleration_{Galaxies} > Gravity_{Universe}$$

Eq 6.5.3: Expansion of Universe Related to Manifest Set Growth Threshold

Conversely:

$$!\,MS\_Growth_{Threshold}\text{: } Acceleration_{Galaxies} < Gravity_{Universe}$$

Eq 6.5.4: Contraction of Universe Related to Manifest Set Growth Threshold

The relation between $MS\_Growth_{Threshold}$ and Dark Energy is suggested by Equation 6.5.5:

$$MS\_Growth_{Threshold} \propto Dark\ Energy$$

Eq 6.5.5: Relation Between Manifested Set Growth Threshold and Dark Energy

In this interpretation so long as more of the infinite set at M_2 results in the manifested-set, MS_{System}, the universe keeps growing.

It is possible to re-figure Figure 6.5.1 with an additional Astronomical Level parallel to the Field Level since the field(s) must act at both the micro and macro scale:

Principle	Field Level	Quantum Level	Atomic Level	Cellular Level	Astronomical Level
System-knowledge	EM Spectrum	Quarks	P-Orbital group	Nucleic Acids	EM Spectrum
System-presence	Mass	Higgs-boson	D-Orbital group	Proteins	Mass
System-power	Energy-gradient implicit in EM Spectrum	Leptons	S-Orbital group	Polysaccharides	Energy-gradient implicit in EM Spectrum
System-nurturing	C, speed of EM Spectrum or Light	Bosons	F-Orbital group	Lipids	C, speed of EM Spectrum or Light

Figure 6.5.2 Order of Magnitude Comparison

In Figure 6.5.2 if the relative distances between levels are considered a pattern emerges. The quantum level is dealing in distances of the order of 10^{-35} m. The cellular level is dealing with distances of the order of 10^{-6} m. The astronomical level is dealing with distances as large as 10^{26} m – the estimated size of the universe. It can be observed therefore that the cell is approximately in the center with roughly 30 orders of magnitude on either side.

Such symmetry that we are at now with the human being at the mid-point between two known material extremes, seems to suggest a cosmic balance of scales with human-will and other faculties such as awareness as a potential arbiter. Lack of will for continued functional-richness will in this view lead to a reversal in the Manifested Set due to a reversal in the quantization trend and consequently to a return of this universe to a state of collapse or unfulfilled absorption into the initiating Light. In that case it will be another possible

Big Bang initiated through the interplay of multiple layers of Light, and a birth following such death as depicted by the ancient Ouroboros symbol of a dragon or snake eating its own tail.

On the other hand the will for continued functional-richness may lead to continuous quantization, space expansion, and a new Ouroboros of eternal return.

A New Ouroboros

Perhaps it is that if the expansion continues the apparent gaps between galaxies filled with extraordinarily high dark-energy becomes ripe for a series of inter-galactic-bangs. And perhaps such a series of inter-galactic-bangs in such a functionally-rich universe precipitates cycles of development based on other speeds of light, that in turn sets up a more conscious fusion of the multiple layers of existence so created, into one incredible Universe where an entirely new type of super-matter can form.

The new Ouroboros will signify a new type of endless creation where richness builds on richness and new types of super-matter on previous super-matter.

The human has in a certain sense been the result of 30 orders of microcosmic magnitude implicit to it. Fields, quanta, atoms, cells, the incredible four-fold emergences of properties codified in Light have all contributed to the emergence of the human. Now in a possible reversal of causality the human appears to be positioned to similarly influence and even structure the further development of the macrocosm by mastery of the microcosm.

Through will and other human faculties continued states of quantum-certainty resulting in four-fold quantization can occur. Such quantization changes the very nature of space, time, energy, and gravity and can alter the development of the 30 orders of macrocosmic magnitude.

Such audacity is only possible because the human is the latest emergence in a universe of Light where all is Light and all is of Light. Light in material form turning on itself can alter the very nature of what it is by what it is. Such is the tale of Ouroboros changing from swallowing its tail to entwining itself through a Cosmos in a never-ending spiral toward unforeseen flowerings.

REFERENCES

1. Arabatzis, T. 2006. Representing Electrons: A Biological Approach to Theoretical Entities. University of Chicago. Chicago
2. Arkani-Hamed, Nima; Bourjaily, Jacob L.; Cachazo, Freddy; Goncharov, Alexander B.; Postnikov, Alexander; Trnka, Jaroslav (2012). "Scattering Amplitudes and the Positive Grassmannian". arXiv:1212.5605
3. Bar-Yam, Y. 2016. From big data to important information. Complexity, 21: 73–98. doi:10.1002/cplx.21785
4. Chown, M. 1990. Can Photons Travel Faster Than Light? New Scientist 126(1711)
5. Cottingham, W & Greenwood, D. 2007. An Introduction to the Standard Model of Particle Physics. Cambridge University Press. Cambridge
6. De Broglie, L. 1929. The Wave Nature of the Electron. Nobel Lecture. http://www.nobelprize.org/nobel_prizes/physics/laureates/1929/broglie-lecture.pdf
7. Deep Order Mathematics Videos. 2016. Deep Order Technologies. http://www.deepordertechnologies.com/new-index
8. Diamond, J. 2005. Collapse: How Societies Choose to Fail or Succeed. Viking Books: New York
9. Einstein, A. 1995. Relativity: The Special and General Theory. New York: Broadway Books.
10. Feinberg, G. 1970. "Particles That Go Faster Than Light". Scientific American. Feb 1970, 69-77.
11. Feynman, RP. 1985. QED The Strange Theory of Light and Matter. New Jersey: Princeton University Press

12. Goodsell, David. 2010. The Machinery of Life. New York: Springer
13. Gottlieb, M. 2013. The Feynman Lectures on Physics: III. California Institute of Technology. http://www.feynmanlectures.caltech.edu/III_16.html
14. Gubser, S. 2010. The Little Book of String Theory. Princeton University Press
15. Harvard-Smithsonian Center for Astrophysics. 2004. https://www.cfa.harvard.edu/seuforum/de_whatmight.htm#
16. Hawking, Stephen. 1988. A Brief History of Time. New York: Bantam Books
17. Heiserman, D. 1991. Exploring Chemical Elements and their Compounds. McGraw-Hill. New York.
18. Hyperphysics. 2016. Department of Physics and Astronomy. Georgia State University. http://hyperphysics.phy-astr.gsu.edu/hbase/mod3.html#c1
19. Isaacson, W. 2008. Einstein: His Life and Universe. Simon and Schuster. New York.
20. Jeans, J. 1932. The Mysterious Universe. Cambridge University Press.
21. Jepsen, K. 2013. Everything is Made of Fields. Symmetry. Joint Publication by Fermilab/SLAC. http://www.symmetrymagazine.org/article/july-2013/real-talk-everything-is-made-of-fields
22. Johnson, S. 2010. Where Good Ideas Come From: The Natural History of Innovation. Riverhead Books: New York
23. Kaufmann, S. 1995. At Home in the Universe. New York: Oxford University Press.
24. Kaufmann, S. 2003. The Adjacent Possible, Edge. https://www.edge.org/conversation/the-adjacent-possible
25. Laszlo, E. 2014. The Self-Actualizing Cosmos: The Akasha Revolution in Science and Human Consciousness. Inner Traditions: Vermont

26. Lorentz, H.A. 1925. The Science of Nature. Vol. 25, p 1008. Springer
27. Malik, P. 2009. Connecting Inner Power with Global Change: The Fractal Ladder. New Delhi: Sage Publications
28. NASA-darkmatter. 2016. http://science.nasa.gov/astrophysics/focus-areas/what-is-dark-energy/
29. NASA/STSci/Ann Field, 2001. Changes in the rate of Expansion Over Time. http://hubblesite.org/image/1037/news_release/2001-09
30. NASA-supernova. 2001. News Release Number: STScI-2001-09: "Blast from the Past: Farthest Supernova Ever Seen Sheds Light on Dark Universe". http://hubblesite.org/newscenter/archive/releases/2001/09/
31. NASA-WMAP, 2014. https://map.gsfc.nasa.gov/media/121238/index.html
32. Ogburn, W. Tomas, D. 1922. Are Inventions Inevitable? A Note on Social Evolution. Political Science Quarterly 37, no.1
33. Olive, K.A et al. 2014. Particle Data Group. Chin. Phys. C, 38, 090001.
34. Panek,R. 2010. Dark Energy: The Biggest Mystery in the Universe. Smithsonian Magazine. http://www.smithsonianmag.com/science-nature/dark-energy-the-biggest-mystery-in-the-universe-9482130/?no-ist
35. Particle Data Group. 2015. Lawrence Berkeley National Laboratory. http://www.cpepphysics.org/main_universe/universe.html
36. Pauli, W. 1964. Nobel Lectures, Physics 1942 – 1962. Elsevier Publishing Company. Amsterdam.
37. Pearson, R. 1997. "Frontier Perspectives", Journal of the Center for Frontier Sciences at

Temple University. Spring/Summer 1997, Volume 6, Number 2, ISBN: 1062-4767.
38. Perkowitz, S. 2011. Slow Light. London: Imperial College Press
39. Planck, M. 1933. Where is Science Going? Ox Bow Press. Connecticut.
40. Portugali, J., 2012. *Self-organization and the city.* New York: Springer Science & Business Media.
41. Prigogine, I. 1977. Time, Structure, and Fluctuations. *Nobelprize.org.* Nobel Media AB 2014. Web. 5 Mar 2016. <http://www.nobelprize.org/nobel_prizes/chemistry/laureates/1977/prigogine-lecture.html>
42. Salkind, N. 2007. Encyclopedia of Measurement and Statistics. Thousand Oaks: Sage Publications.
43. Snyder, M. 2010. Stanford Medicine. http://med.stanford.edu/news/all-news/2010/03/what-makes-you-unique-not-genes-so-much-as-surrounding-sequences-says-stanford-study.html#.html
44. Sri Aurobindo. 1971. Social and Political Thought. Sri Aurobindo Ashram Press: Pondicherry
45. Stewart, Ian. 2012. In Pursuit of the Unknown. Basic Books. New York.
46. Thomas, D, Brown, J. 2009. Why Virtual Worlds Can Matter. International Journal of Learning and Media 2009 1:1, 37-49
47. Toynbee, A. 1961. A Study of History, Volumes I – XII. Oxford University Press: Oxford
48. Turom, N. 2012. The Universe Within. Anasi Press: Ontario
49. Tweed, M. 2003. Essential Elements: Atoms, Quarks, and the Periodic Table. Walker & Copmany. New York.
50. Usó -Doménech, J, Nescolarde-Selva, J, Lloret-Climent, M, Fan, M. 2016. Semiotic open complex systems: Processes and behaviors. Complexity, 21: 388–396. doi:10.1002/cplx.21817

51. Whitaker, A._2006. Einstein, Bohr and the Quantum Dilemma: From Quantum Theory to Quantum Information. Cambridge: Cambridge University Press
52. Wilczek, F. 2016. A Beautiful Question: Finding Nature's Deep Design. New York: Penguin Books
53. Wolchover, N. 2013. A Jewel at the Heart of Quantum Physics. Quanta Magazine . https://www.quantamagazine.org/20130917-a-jewel-at-the-heart-of-quantum-physics/
54. Wright, R. 2009. Evolution of Compassion. https://www.ted.com/talks/robert_wright_the_evolution_of_compassion/transcript?language=en. TED 2009.
55. Yates, F.E. 2012. *Self-organizing systems: The emergence of order*. New York: Springer Science & Business Media.

Made in the USA
Monee, IL
05 November 2026

67386702R10111